Petr Filip and Martin Pivokonský
Coagulation and Flocculation

Also of interest

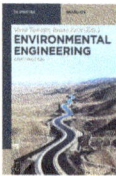

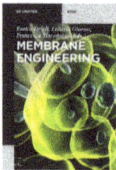

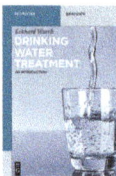

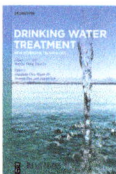

Petr Filip and Martin Pivokonský

Coagulation and Flocculation

In Drinking Water Treatment

DE GRUYTER

Authors

Dr. Petr Filip
Department of Hydrochemistry and Technology
Institute of Hydrodynamics
Czech Academy of Sciences
Pod Paťankou 30/5
160 00 Prague, Czech Republic
filip@ih.cas.cz

Dr. Martin Pivokonský
Department of Hydrochemistry and Technology
Institute of Hydrodynamics
Czech Academy of Sciences
Pod Paťankou 30/5
160 00 Prague, Czech Republic
pivo@ih.cas.cz

ISBN 978-3-11-124597-3
e-ISBN (PDF) 978-3-11-124676-5
e-ISBN (EPUB) 978-3-11-124721-2

Library of Congress Control Number: 2025939257

Bibliographic information published by the Deutsche Nationalbibliothek
The Deutsche Nationalbibliothek lists this publication in the Deutsche Nationalbibliografie;
detailed bibliographic data are available on the internet at http://dnb.dnb.de.

© 2025 Walter de Gruyter GmbH, Berlin/Boston, Genthiner Straße 13, 10785 Berlin
Cover image: Martin Pivokonský, Institute of Hydrodynamics of the CAS
Typesetting: Integra Software Services Pvt. Ltd.

www.degruyterbrill.com
Questions about General Product Safety Regulation:
productsafety@degruyterbrill.com

Preface

The field of water treatment encompasses a diverse array of topics, spanning from the atomic scale – covering ions, repulsive, and attractive forces – to the large-scale infrastructure of sedimentation tanks and filters. This breadth is spanned with theoretical models, experimental measurements, and practical applications within the discipline. Adding to the complexity, the raw water characteristics vary significantly across individual water treatment plants (WTPs) and are seldom consistent throughout the year. Factors such as seasonal variations, climate change, and societal development further contribute to fluctuations in water quality. Financial considerations also play a critical role in determining treatment approaches. Recent years have seen a growing trend towards replacing traditional materials used in the treatment process with environmentally friendly alternatives. Nevertheless, the final product – drinking water – must be compliant meeting strict qualitative standards.

This book aims to provide a concise introduction to the coagulation/flocculation process, presented across four independent chapters. Each chapter is designed to stand alone, allowing the reader to explore topics independently. The book begins with an introduction to the forces acting on impurities (particles) and the initiation of destabilisation, including particle clustering. It covers the classical interplay of factors such as coagulant selection, dosage, and optimal pH value. The final chapter focuses on particle compactness – fractal dimension – and explores its theoretical and experimental determination, as well as the implications for aggregate breakage and sedimentation.

The contents of this book are designed to be accessible to a broad audience, whether from a research or practical background. Mathematical complexity is minimised, but where simple derivations are possible, they are included.

P.F., M.P.
Prague, Czech Republic
March 31, 2025

Acknowledgement: The authors appreciate the support of The Czech Academy of Sciences Premium Academiae.

https://doi.org/10.1515/9783111246765-202

Contents

Preface —— V

Chapter 1
Particles and their interaction —— 1
1.1   Particle specification —— 1
1.2   Basic characterisation of particles —— 2
     1.2.1   Brownian motion —— 2
     1.2.2   Geometrical description of particles —— 5
1.3   Colloidal particles —— 9
1.4   Van der Waals forces —— 10
     1.4.1   Van der Waals attraction energy —— 11
     1.4.2   A determination of the Hamaker constant —— 18
     1.4.3   Rate of the attractive, drag, and gravitational forces —— 20
1.5   Lennard-Jones potential —— 22
1.6   Electrical double layer —— 24
1.7   The Derjaguin-Landau-Verwey-Overbeek (DLVO) theory —— 27
1.8   The eXtended DLVO (XDLVO) theory —— 28
     1.8.1   Lewis acid-base interaction —— 29
     1.8.2   Born repulsion —— 31
     1.8.3   Solvation —— 32
     1.8.4   Hydrophobic attraction —— 32
     1.8.5   Steric interaction —— 34
     1.8.6   Polymer bridging —— 35
     1.8.7   Depletion —— 36
References —— 38

Chapter 2
Destabilisation of colloid systems – its initiation —— 43
2.1   Destabilisation of colloid systems – its evaluation —— 44
     2.1.1   Zeta (ζ) potential —— 44
     2.1.2   Water density distribution function —— 45
     2.1.3   Temperature effects on a rate of coagulation —— 47
     2.1.4   Electrokinetic phenomena —— 49
         2.1.4.1   Electroosmotic flow —— 50
         2.1.4.2   Electrophoretic flow —— 55
         2.1.4.3   Streaming potential —— 56
         2.1.4.4   Sedimentation potential —— 58
     2.1.5   Characteristics of the total interaction energy curve —— 59
     2.1.6   Correctness of the ζ-potential evaluation —— 61
2.2   Schulze-Hardy and inverse Schulze-Hardy rules —— 63

2.2.1   Classical Schulze-Hardy rule —— **63**
2.2.2   Modified Schulze-Hardy rule —— **65**
2.2.3   Inverse Schulze-Hardy rule —— **67**
2.3   Coagulants – basic categorisation —— **68**
2.4   Coagulant dosage —— **70**
2.4.1   Jar test —— **70**
2.4.2   Artificial intelligence techniques —— **73**
2.4.3   Regression analysis —— **76**
References —— **77**

Chapter 3
Process of drinking water treatment using coagulation/flocculation
method —— 84
3.1   Density of measurements – pH vs. concentration of ions H^+ —— **84**
3.2   Natural organic matters (NOMs) —— **87**
3.3   Hydrolysis of metal cations —— **88**
3.4   An influence of increasing coagulant/flocculant dosage on water
        characteristics —— **91**
3.5   Basic aggregation mechanisms —— **93**
3.6   Traditional frequently used coagulants —— **93**
3.7   Response surface method —— **100**
3.8   Impact of societal development on water treatment procedures —— **105**
3.8.1   Algal organic matters (AOMs) —— **105**
3.8.2   Pre-oxidation processes —— **106**
3.8.3   Chlor(am)ination and disinfection by-products (DBP) —— **106**
3.9   Mechanical aspects of aggregation —— **108**
References —— **111**

Chapter 4
Fractal nature of aggregates, experimental determination —— 116
4.1   Introduction to a fractal dimension —— **116**
4.2   Application of a fractal dimension to coagulation and flocculation —— **124**
4.3   Evaluation of a fractal dimension —— **132**
4.4   Determination of a fractal dimension —— **138**
4.4.1   Initial fractal approaches —— **138**
4.4.1.1   Box counting method —— **138**
4.4.1.2   Sand box technique – nesting square method —— **142**
4.4.2   Electrical and optical methods —— **142**
4.4.2.1   Electrical methods —— **143**
4.4.2.2   Optical methods used in fractal analysis —— **144**
4.4.3   Fractal dimension based on settling velocity —— **164**
4.4.4   Fractal dimension in relation to pH —— **168**

4.4.5   Fractal dimension in a relation to temperature —— **171**
4.4.6   Fractal dimension vs coagulants —— **172**
4.4.7   Computer modelling of fractal aggregates —— **174**
4.4.7.1   Reaction-limited particle-cluster aggregation model (RLA) —— **176**
4.4.7.2   Ballistic particle-cluster aggregation model (BA) —— **176**
4.4.7.3   Diffusion-limited particle-cluster aggregation model (DLA) —— **177**
4.4.7.4   Summary of all three models (1)–(3) —— **178**
4.5   Summary of fractal approaches —— **181**
References —— **182**

List of abbreviations —— **191**

Index —— **193**

Chapter 1
Particles and their interaction

The particle represents a crucial term in the process of water treatment and – in principle – of similar technologies. To eliminate their unwanted presence, it is not possible to restrict the approaches to the physical procedures and mechanical means only as the sizes of the particles are not bounded from below. For the dimensions below micrometres and approaching to nanometres, it is necessary first – using a chemical way – to destabilise the particles, then to force them to aggregate, and consequently as a final step to remove them from a carrier liquid. This procedure is discussed in the following chapters. This introductory chapter is devoted to particle characterisation and attractive and repulsive forces which a particle is exposed to.

1.1 Particle specification

A word (solid) particle represents a principal term in the processes of coagulation and flocculation. This term can be understood at three more or less distinct positions:
1) The attributes of particles in practice: various usually non-spherical shapes, rough surface, various particles created from various materials, porosity.
2) The assumptions required by the basic classical approaches modelling particles behaviour: monodisperse spherical particles, smooth surface, identical uniformly distributed material with an identical refractive index, non-porous character.
3) An acceptable balance between practice (e.g. the particle characteristics in the water treatment plants) and necessary simplifications in the passage from (2) to (1) enabling a theoretical development of new procedures using contemporary experimental and computational techniques. This point is really crucial and does not concern only theoretical modelling but also a suitable generalisation of the purely experimental conclusions.

The terminology of the coagulation and flocculation processes also faces some commercial trends as, for instance, the term 'nanoparticles', which is now used even in the micron dimensions. Nevertheless, its categorisation can be simply based on the appearance of the individual forces acting on the particles and their comparison. In each category some forces are dominant and the other ones can be neglected. As a gravitational force has no impact on forces balancing on atomic level, an impact of the van der Waals forces (distance-dependent interaction between atoms or molecules) strongly attenuates in the case when the particles lose their nanomaterial character, i.e., if their dimensions exceed tens of nanometres.

https://doi.org/10.1515/9783111246765-001

1.2 Basic characterisation of particles

First, a dependence of Brownian motion on temperature will be specified indicating non-negligible changes in particle behaviour. Consequently, a boundary between colloidal and suspended particles will be determined including numerical justification. It will be shown that a usually fixed value (boundary) 1 μm subjects to the material studied and may be substantially different.

1.2.1 Brownian motion

Brownian motion denotes a phenomenon when small particles are exposed to permanent random moderate fluctuations. If no direction of the random oscillations is preferred, then particle distribution will be balanced throughout the medium over a period of time; this process is called diffusion. The Brownian motion of the small particles (lower than microns) is the result of accumulated kinetic energy in molecules of a carrier liquid, which initiates their continuous chaotic motion and permanent pushing of the particles by the molecules with simultaneous imparting of kinetic energy.

The concentration of Brownian particles φ at point x at time t satisfies the diffusion equation:

$$\frac{\partial \varphi}{\partial t} = D \cdot \frac{\partial^2 \varphi}{\partial x^2},\tag{1.1}$$

where the diffusion coefficient D relates the diffusive flux with the changes in concentration gradient (Fick's second law).

Denoting N as a number of particles at the initial time $t = 0$, the diffusion equation has the analytical solution

$$\varphi(x,t) = \frac{N}{\sqrt{4\pi Dt}} \cdot e^{-\frac{x^2}{4Dt}}.\tag{1.2}$$

This expression is identical with a normal distribution (see Fig. 1.1) in its standard form:

$$f(x) = \frac{1}{\sigma\sqrt{2\pi}} \cdot e^{-\frac{(x-\mu)^2}{2\sigma^2}}$$

with the mean $\mu = 0$ and variance $\sigma^2 = 2Dt$ (σ is the standard deviation).

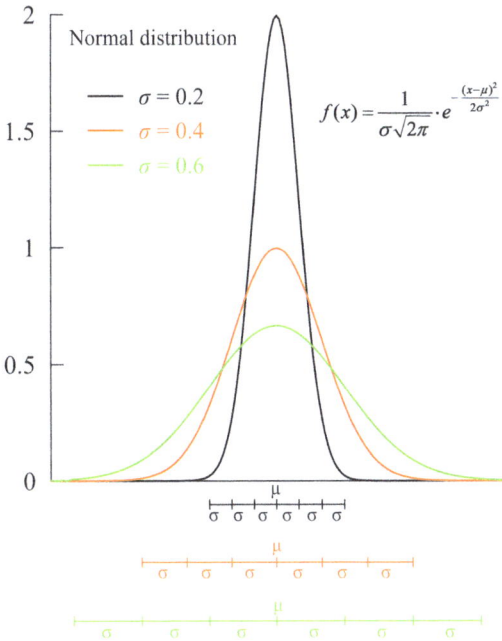

Fig. 1.1: The bell-shape courses of the normal distribution for which 68.27% values are located in the interval $[\mu-\sigma, \mu+\sigma]$, 95.45% in $[\mu-2\sigma, \mu+2\sigma]$, and 99.73% in $[\mu-3\sigma, \mu+3\sigma]$.

In other words, the first central moment of this distribution is 0, which implies the same probability whether a particle will move to the left or to the right. The non-vanishing second central moment (equal to variance σ^2)

$$\overline{x^2} = 2Dt \tag{1.3}$$

provides the first approximation for an estimate of the mean displacement $\bar{x}$ of the particle within the time interval $[0, t]$:

$$\bar{x}^2 = 2Dt \implies \bar{x} = \sqrt{2Dt}. \tag{1.4}$$

For the spherical particles of a fixed diameter d and low non-dimensional Reynolds number Re (relating the inertial and viscous forces), the Stokes-Einstein-Sutherland equation expresses the diffusion coefficient D in the form relating thermal energy of the particle (a numerator) with the drag experienced by the particle (a denominator):

$$D = \frac{k_B T}{3\pi d\eta}, \tag{1.5}$$

where k_B is the Boltzmann constant (as part of the 2019 redefinition of SI base units the Boltzmann constant is defined to be exactly 1.380649×10^{-23} J·K^{-1} (= kg·m^2·s^{-2}· K^{-1})), T is the absolute temperature in K, and η is the dynamic viscosity in Pa·s (= kg·m^{-1}·s^{-1}).

By inserting eq. (1.5) into eq. (1.4) we can obtain an estimate for the mean displacement of the particles; for $d = 0.5$ and 1 μm, see Tab. 1.1 (presented for water). The absolute accuracy indicated in the column 'Mean displacement' is within a rough estimate (two decimal figures are out of reality). However, it provides a comparison of mean displacements between various temperatures. We can see from Tab. 1.1 that a ratio of mean displacement at temperatures 10 and 20 °C attains 0.86.

Tab. 1.1: An estimate of mean displacement of colloid particles in water at various temperatures.

Diameter (μm)	Temperature (°C)	Dynamic viscosity (Pa · s)	Time (s)	Diffusion coefficient (10^{-13} m^2 · s^{-1})	Mean displacement (μm)
0.5	10	0.0013059	1	6.35	1.13
0.5	15	0.0011375	1	7.42	1.22
0.5	20	0.0010016	1	8.58	1.31
0.5	10	0.0013059	2	6.35	1.59
0.5	15	0.0011375	2	7.42	1.72
0.5	20	0.0010016	2	8.58	1.85
1	10	0.0013059	1	3.18	0.80
1	15	0.0011375	1	3.71	0.86
1	20	0.0010016	1	4.29	0.93
1	10	0.0013059	2	3.18	1.13
1	15	0.0011375	2	3.71	1.22
1	20	0.0010016	2	4.29	1.31

Using the semi-empirical Vogel-Fulcher-Tammann relation for approximating water viscosity (in Pa·s)

$$\eta(T) = 0.00002939 \times \exp\left(\frac{507.88}{T - 149.3}\right) \tag{1.6}$$

and using the expression derived from eqs. (1.4) and (1.5) which relates the mean displacements at various temperatures T_1 and T_2:

$$\text{Ratio}_{\text{displ}}(T_1, T_2) = \sqrt{\frac{T_1 \cdot \eta(T_2)}{T_2 \cdot \eta(T_1)}}, \tag{1.7}$$

we obtain the time-invariant increasing curve (see Fig. 1.2), where $T_2 = 283.15$ K (= 10 °C) was taken as a reference temperature. From here we can see that a mean displacement in water at 25 °C in comparison with 4 °C is increased by more than one third; for a couple 10 (frequent temperature in the water treatment plants) and 25 °C (laboratory conditions) an increase attains one quarter. The course of eq. (1.7) with $T_2 = 283.15$ K (= 10 °C) can be approximated (practically interpolated) by a straight line:

$$\text{Ratio}_{\text{displ}}(Temp) = 0.84 + 0.016 \times Temp, \tag{1.8}$$

where temperature $Temp$ is in °C.

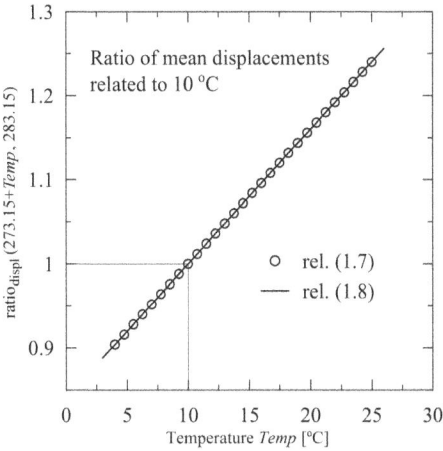

Fig. 1.2: The time-invariant relative mean displacements of particles at various temperatures related to the reference displacement at 10 °C.

1.2.2 Geometrical description of particles

The spherical shape of particles representing an absolute symmetry is accompanied by a series of other very pleasant (and simplifying) attributes as apparent from Fig. 1.3, where a particle is substituted by its spherical (volumetric) equivalent:

- the particle morphology can be characterised by a single parameter (a diameter),
- rough surface is replaced by smooth surface,
- as the volumes should be identical, porosity is eliminated,
- there is only one point where two spherical particles mutually touch, and two non-spherical particles can exhibit multipoint touch (including faces),
- optical properties are unambiguously defined,
- resistance against flow (drag force) is constant regardless of the mutual position between flow direction and a particle inclination,
- two-dimensional (2D) projection of a sphere is an identical circle regardless of a particle position and a direction of projection,
- a centre of rotation, if a sphere is sheared, is always at the centre of sphere strictly contradicting to a non-spherical particle, a drag force can be simply interpreted,
- a ratio 'volume/surface' is fixed for a spherical particle (=one sixth of a diameter).

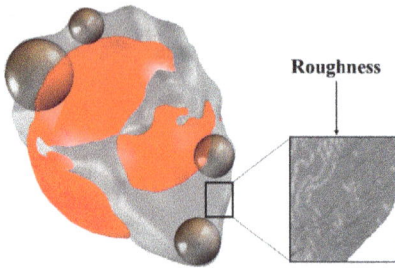

Fig. 1.3: An equivalent (red) sphere occupying the same volume as a particle including a change in surface roughness (taken from Khan and Latha 2023, Fig. 6).

As can be deduced from above, a complexity of the topic 'coagulation and flocculation' when general particles are taken into account is very complicated and generally impossible to solve. This is a reason why a series of proposals have appeared trying to describe non-spherical particles in a more descriptive way than only as simplified projected circles and spheres.

Systematic analysis of non-spherical particle characterisation started in the 1920s and was based on the so-called tri-dimensional approach L, I, and S. The length L is assigned to the longest dimension of the particle, I represents the longest dimension perpendicular to L, and S is the longest dimension perpendicular to both L and I. Nevertheless, a determination of these three relatively simply defined lengths is not easy in practice and not always their defined determination is the optimal one. This can be illustrated by an example of a cube, where the longest dimension is given by its diagonal connecting the opposite corners. A natural choice is to consider the individual edges ($L = I = S$).

Various form descriptors were gradually introduced using various L-I-S combinations such as Wentworth flatness index $(L + I)/(2\,S)$ (Wentworth 1922), Krumbein intercept sphericity $\sqrt[3]{IS/L^2}$ (Krumbein 1941), Corey shape factor $S/\sqrt{LI}$ (Corey 1949), elongation S/I and flatness I/L (Zingg 1935), and other more elaborated factors. The common denominator of all formulas is their smooth passage to a parameter characterising a sphere; in this case, a value of 1. Application of inscribed and circumscribed circles and spheres resulting in a definition of circularity, roundness, and sphericity is indicated in Fig. 1.3. For illustration, Wadell (1932) introduced a roundness as a mean of N individual corner geometries ($i = 1, \ldots, n$) in the form $\sum (r_i/R)/N$, where r_i and R are the radius of curvature of a corner and the radius of maximum inscribed circle in the analysed plane, respectively. More details including recent literature is presented in Blott and Pye (2008) and Anusree and Latha (2023).

In the case of uniform particles (the same size, shape, or mass) or nearly uniform, an application of the suitable chosen descriptor has its substantiation. However, in the case of strictly non-uniform particles (an inconsistent size, shape, and mass distribution) an introduction of any descriptor may be questioned.

Loosely speaking, importance of particle shape attenuates with its size. To substantiate this assertion, let us compare the time intervals (t_{diff} and t_{set} in diffusion and settling regimes, respectively) during which a spherical particle will travel a distance equal to its diameter d.

For a calculation of t_{diff}, a combination of eqs. (1.4) and (1.5) is used:

$$x = \sqrt{\frac{2k_B T}{3\pi d\eta}} \cdot t. \tag{1.9}$$

As $x = d$ for t_{diff} we get

$$t_{diff} = \frac{3\pi\eta}{2k_B T} \times d^3. \tag{1.10}$$

Dynamic viscosity of water is $\eta = 0.0010016$ Pa.s for $T = 293.15$ K (20 °C). Inserting these values, we obtain an approximate relation

$$t_{diff} = 1.166 \times 10^{18}\, d^3. \tag{1.11}$$

For a calculation of t_{set} we use the Stokes equation (for details, see Chapter 4) determining a settling velocity v_{set}:

$$v_{set} = \frac{(\rho_p - \rho_l)g}{18\eta} \times d^2, \tag{1.12}$$

where $\rho_p - \rho_l$ is the difference between particle and carrier liquid (water) densities, g is the gravitational acceleration (ranging from 9.764 to 9.834 m·s^{-2} depending on altitude, latitude, and longitude, a conventional standard value is defined exactly as 9.80665 m·s^{-2}). Multiplying both sides by t_{set} we get ($v_{set} \times t_{set} = d$)

$$t_{set} = \frac{18\eta}{(\rho_p - \rho_l)g} \times d^{-1}.$$

If we consider kaolinite particles ($\rho_p = 2{,}650$ kg·m^{-3}) we obtain

$$t_{set} = 1.114 \times 10^{-6}\, d^{-1}. \tag{1.13}$$

Relating eqs. (1.11) and (1.13) we get an approximate diameter for which the diffusion and sedimentation times are equal:

$$d^4 = 0.955 \times 10^{-24} \implies d = 0.99 \times 10^{-6}\ [\text{m}] \implies d \approx 1\,\mu\text{m}. \tag{1.14}$$

This rough calculation explains why, by convention, the colloidal particles – for which diffusion dominates settling – are bounded from above by a value of approximately 1 μm.

A comparison of various particles is illustrated in Fig. 1.4, where the classification of clay, silt, and sand particles is after ISO 14588-1 (2002). The sequence 0.2, 0.63, 2, 6.3, ... is not accidental but in the logarithmic scale corresponds to an equidistant sequence −0.7, −0.2, 0.3, 0.8, ... with the difference 0.5 between the consequent members. The graphically indicated ranges for other particles represent the frequently reported data in the literature not involving limiting cases.

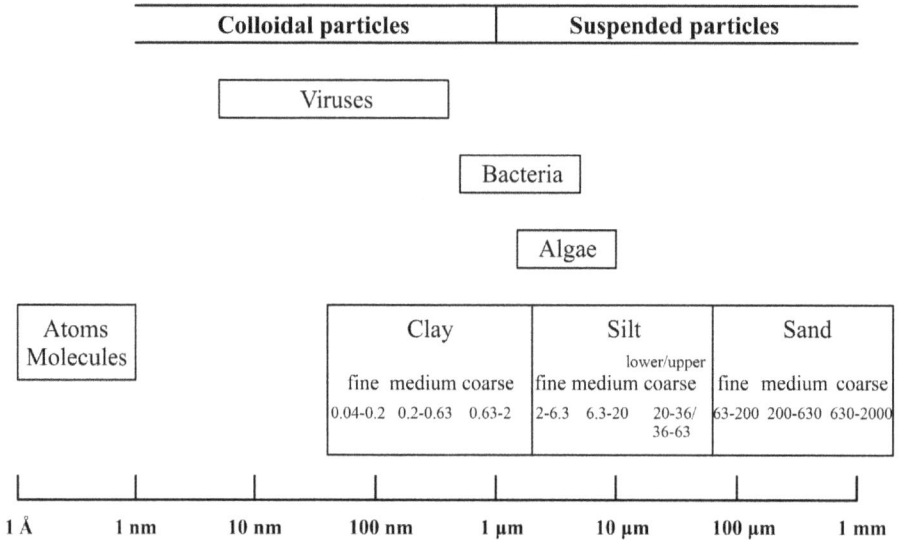

Colloidal particles		Suspended particles	

	Viruses		

		Bacteria	

		Algae	

Atoms Molecules	Clay	Silt	Sand
		lower/upper	
	fine medium coarse	fine medium coarse	fine medium coarse
	0.04-0.2 0.2-0.63 0.63-2	2-6.3 6.3-20 20-36/ 36-63	63-200 200-630 630-2000

1 Å   1 nm   10 nm   100 nm   1 µm   10 µm   100 µm   1 mm

Fig. 1.4: The ranges of frequently encountered particles, the individual ranges are given in micrometres ($1\,\text{Å} = 10^{-10}$ m).

The boundary (1 µm) separating colloidal and suspended particle is necessary to consider as a generally accepted value. In Fig. 1.4 the upper limit of algae range attains a value of 10 µm (exceeding by one order the separating value 1 µm). However, if we use the density $\rho_p = 1.1$ kg·m^{-3} corresponding to algae, then applying the procedure analogous to that above we receive $d \approx 11$ µm, a diameter for which the diffusion and sedimentation times are balanced.

A geometrical shape of colloidal particles is not as determinative as for the particles for which settling dominates and are closely connected with a drag force. This is connected with a ratio surface/volume. If we consider a spherical colloidal particle, then we get

$$\frac{\text{Surface}}{\text{Volume}} = \frac{\pi d^2}{\pi d^3/6} = \frac{6}{d}. \tag{1.15}$$

For a cube with an edge a we obtain an analogous result $6/a$, and for a tetrahedron $6\sqrt{6}/a$. As we can see these ratios (for a, $d < 10^{-6}$) attain very large values (and not only for regular particles). This is a reason why a detailed knowledge of surface characteristics of colloidal particles is a crucial point for analysing their behaviour.

1.3 Colloidal particles

Generally, colloids are classified according to the phases of dispersed substance and media in which they are distributed. There are four basic groups:
-   Sol: solid particles are dispersed in a liquid;
-   Emulsion: both materials are liquids;
-   Foam: gas particles dispersed in a liquid or solid;
-   Aerosol: liquid or solid particles dispersed in a gas.

The attention will be prevailingly paid to the first group: sols as the most studied topic in water treatment. However, a flotation process uses air bubbles for buoying the unwanted impurities and their elimination.

It is not trivial to describe a mutual interaction of colloidal particles as the distances between the individual objects range from approximately 1 Å ($=10^{-10}$ m) roughly corresponding to an atom size up to about 1 μm (10^{-6} m) depending on the material studied (see above). From the viewpoint of the classical mechanics it spreads from the balance equations with no impact of the gravitational acceleration up to its initial consideration in a passage to a suspension flow. Among the inputs contributing to a description of particles behaviour, for illustration, it is possible to pick up these types of colloid interaction (Gregory 2006):
-   van der Waals (usually attractive);
-   Electrical double layer (either repulsive or attractive);
-   Hydration effects (repulsive);
-   Steric interaction of adsorbed layers (usually repulsive);
-   Polymer bridging (attractive).

By far this list is not complete and other types of colloid interactions participate in the overall balancing (introduced later), among other things strongly influencing and eliminating otherwise non-adequate asymptotic behaviour of the introduced five types of interaction (e.g. van der Waals forces vs. Born forces in a close (molecular) vicinity of surfaces).

1.4 Van der Waals forces

The so-called van der Waals forces represent one group of the intermolecular forces. They originate from the dipole interactions at the atomic level, and hence, their importance in the colloidal structure increases with the presence of particles which are constituted by a large number of atoms and molecules.

An electric dipole is created by the separation of the positive and negative electric charges. A measure of these charges within a system is called the electric dipole moment with the SI unit Cm (Coulomb-metre); in the CGS system, a unit is denoted Debye (the first researcher who extensively studied molecular dipoles). In the case of molecules, three different types of dipoles are distinguished:

a) permanent dipoles: they appear in the case that two atoms in a molecule exhibit substantially different electronegativity. This results in attracting more electrons by one atom and hence, more positivity of the other one.

b) instantaneous dipoles: their occurrence is characterised by higher concentration of electrons in one place of a molecule. These temporary dipoles are of lower magnitude than the permanent ones.

c) induced dipoles: these dipoles are evoked by a permanent dipole in a neighbouring molecule which repels the electrons and in such a way induces polarity (an induced dipole).

The van der Waals forces are composed from three types of the inputs (a historical review is introduced in Israelachvili and Ninham (1977)):

– the Keesom force (Keesom 1915) between permanent molecular dipoles in which rotational orientations are dynamically averaged over time (hence a term orientation energy); the Keesom forces are temperature-dependent and energy is proportional to the inverse sixth power of the distance. No Keesom interactions are present in aqueous solutions containing electrolytes.

– the Debye force (Debye 1920) arises from interactions between rotating permanent dipoles and induced dipoles (hence a term induction energy). The Debye forces cannot occur between atoms and are much less temperature-dependent than the Keesom forces but with the same power decrease with the distance.

– the London force (London 1930) (induced dipoles-induced dipoles) is the result of random fluctuations of electron density in an electron cloud. These forces are also denoted as the dispersion forces (terminology derived from light dispersion; Mahanty and Ninham 1976) and again decrease with the same power as the preceding ones.

The London interactions are the most important contributions to the van der Waals forces in comparison with the Keesom and Debye forces as they are based on ever-present material polarisability and not requiring the presence of the permanent dipoles as the remaining two forces. In some materials the London forces dominate

(CCl$_4$, benzene – 100%). For some materials the participation (Keesom, Debye, London) is more balanced (ethanol – 42.6, 9.7, 47.6) and a completely diverse situation is exhibited in water (84.8, 4.5, 10.5) due to small and highly polar molecules. [The data are taken over from Hiemenz and Rajagopalan (1997, p. 477) and McClellan (1963), the deviations from the exact 100% in summation are caused by rounding off.] From the principle of their introduction, the van der Waals forces are anisotropic because they depend on the relative orientation of the molecules. The so-called retardation effect is connected with the London dispersion forces when two atoms are an appreciable distance apart. This causes that the direction of the instantaneous dipole of the first atom after its reaching the second atom and return completely differs from its initial orientation. In contrast to the London energy, the orientation and induced energies are not exposed to the retardation effects (Israelachvili 2011, p. 130).

The van der Waals attractive intermolecular forces are not based on a chemical electronic bond as the covalent or ionic bonds and in a comparison with these bonds are lower. The van der Waals forces also remarkably drop with an increasing distance between interacting molecules (particles).

With a decreasing distance between the atoms and absence of other forces, the van der Waals force gradually changes its characterisation from the attractive to rather repulsive manifestation as a result of the mutual repulsion between the atoms' electron clouds. The separating point is denoted as the van der Waals contact distance.

1.4.1 Van der Waals attraction energy

Hamaker (1937) extended the van der Waals interactions from molecules to large bodies. In this sub-section an emphasis is paid to mutual interaction of two generally not equisized spherical particles. Other geometries such as sphere-plane, rectangle-rectangle, parallel and perpendicular cylinders, and many others are tabulated in Parsegian (2006).

The van der Waals attraction energy of two generally unequal spherical particles with the radii r_1 and r_2 are expressed by the relation (Hamaker 1937, eqs. (13, 13a); Parsegian 2006, p. 155):

$$V_{\mathrm{a}}(L;r_1,r_2) = -\frac{A}{6}\left[\frac{2r_1r_2}{L^2-(r_1+r_2)^2}+\frac{2r_1r_2}{L^2-(r_1-r_2)^2}+\ln\left(\frac{L^2-(r_1+r_2)^2}{L^2-(r_1-r_2)^2}\right)\right], \qquad (1.16)$$

where L is the distance between both centres and A is the Hamaker constant representing the coefficient of the interaction between large objects (Parsegian 2006). This relation was calculated by Hamaker (1937) from the relation expressing the interaction energy between two spherical particles with a number density ρ (atoms per cm^3):

$$V_a(L;r_1,r_2) = -\int_{V_1}\left(\int_{V_2}\frac{\lambda\rho^2}{r_d^6}\,dv_2\right)dv_1, \tag{1.17}$$

where V_1 and V_2 are the volumes of the first and second particles, respectively; dv_1 and dv_2 are the volume elements with the distance r_d; λ is the London-van der Waals constant.

The (nowadays denoted) Hamaker constant A is expressed as

$$A = \lambda\pi^2\rho^2. \tag{1.18}$$

For particles of different materials (with number densities ρ_1 and ρ_2) the Hamaker constant is of the form

$$A = \lambda\pi^2\rho_1\rho_2. \tag{1.19}$$

The values of the Hamaker constants are roughly in the range $1\times10^{-19} \div 1\times10^{-20}$ (Hiemenz and Rajagopalan 1997, p. 485; Israelachvili 2011, pp. 253–254). Specifically, for instance, the Hamaker constants for water and Al_2O_3 in vacuum attain the values 4.35×10^{-20} J and 15.4×10^{-20} J, respectively (Bargeman and van Voorst Vader 1972).

Denoting

$$L = r_1+r_2+l, \quad l_r = \frac{l}{r_1}, \quad q_r = \frac{r_2}{r_1}, \tag{1.20}$$

where l represents the distance between the spherical particles (a clearance, not between the centres), we can re-write eq. (1.16) to the form

$$V_a(l_r,q_r) = -\frac{A}{6}\left[\frac{2q_r}{(1+q_r+l_r)^2-(1+q_r)^2}+\frac{2q_r}{(1+q_r+l_r)^2-(1-q_r)^2}\right.$$
$$\left.+\ln\left(\frac{(1+q_r+l_r)^2-(1+q_r)^2}{(1+q_r+l_r)^2-(1-q_r)^2}\right)\right]. \tag{1.21}$$

Here we can see that the Hamaker constant A is a coefficient linearly relating the interactive van der Waals energy to the distance of separation between two molecules where the interactive force is pair-wise additive (dv_1 and dv_2 in the integral (1.17)) and independent of the intervening media.

If we fix r_1 and choose two different values for the particles distance (e.g. $l_r = 0.1$ and 0.3, one and three tenths of a radius r_1), we can see dramatic decrease in van der Waals attraction energy for a series of radii of the other spherical particle (see Fig. 1.5).

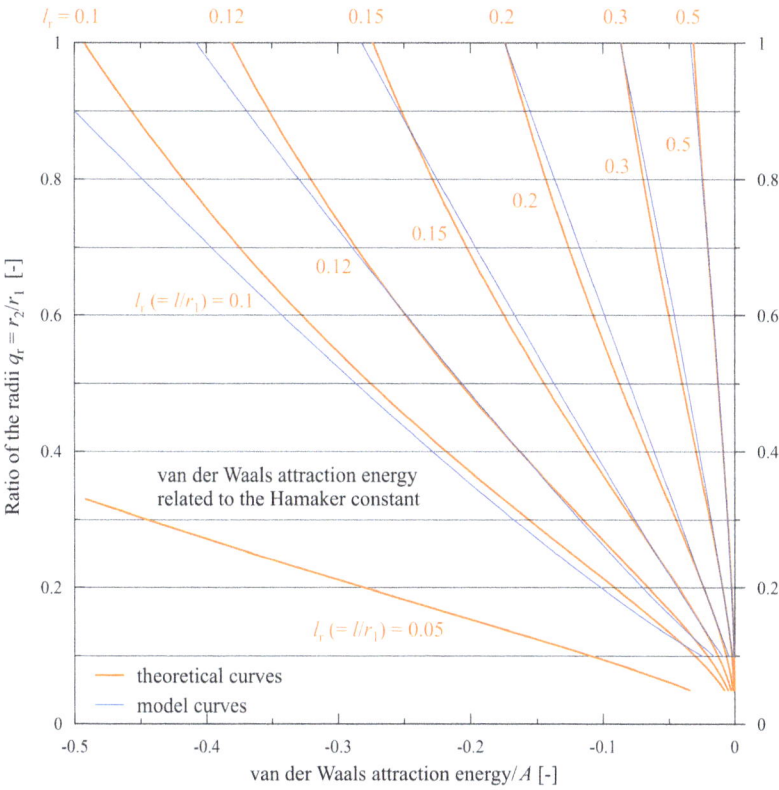

Fig. 1.5: Dependence of the van der Waals attraction energy related to the Hamaker constant on the quotients q_r and l_r relating basic geometrical inputs of two spherical particles (two radii and particle distance). For better legibility the final portions of the curves with constants l_r approaching the point [0,0] are not depicted.

For practical application not always the precise expressions are necessary and the approximate relations with acceptable deviations can be useful (see Bowen and Jenner, 1995).

The algebraically rather complicated eq. (1.21) can be approximated (for $q_r \geq 0.1$ and $l_r \geq 0.1$) by

$$V_a^{appr}(l_r, q_r) = A \cdot a(l_r) \cdot [q_r - b(l_r)]^{c(l_r)}, \tag{1.22}$$

where

$$a(l_r) = 0.00843 - 0.015 \cdot l_r^{-1.6}, \quad b(l_r) = 0.0787 - 0.0135 \cdot l_r^{1.26}, \quad c(l_r) = 0.326 + 1.28 \cdot l_r^{0.4}. \tag{1.22a, b, c}$$

The approximations – depicted in blue in Fig. 1.5 – are, for instance, for the values $l_r = 0.12$, 0.2, and 0.3 of the form:

$$V_a^{appr}(0.12, q_r) = -0.438 \cdot [q_r - 0.0778]^{0.874} \times A, \tag{1.23a}$$

$$V_a^{appr}(0.20, q_r) = -0.186 \cdot [q_r - 0.0769]^{0.998} \times A, \tag{1.23b}$$

$$V_a^{appr}(0.30, q_r) = -0.945 \cdot [q_r - 0.0757]^{1.117} \times A. \tag{1.23c}$$

This provides much better basic orientation in the functional changes of the van der Waals attraction energy in dependence on the mutual relation between the sizes of the particles than eq. (1.21).

For the limiting case, when the radii of both spherical particles are identical ($\equiv r$), eq. (1.16) can be simplified to the form (Stumm and Morgan 1996, p. 868):

$$V_a(s) = -\frac{A}{6}\left[\frac{2}{s^2 - 4} + \frac{2}{s^2} + \ln\left(\frac{s^2 - 4}{s^2}\right)\right], \tag{1.24}$$

where $s = L/r$ ($\equiv 2 + l/r$). Figure 1.6 documents strong attenuation of the van der Waals attraction energy with a relatively small change of the particles' distance. The passage from a distance representing one tenth of the particle radius to a distance where approximately one more particle can be placed between the analysed particles represents a decrease by three orders, i.e., energy in the latter case is 0.1% of the former energy. This drop is invariant with respect to a value of the Hamaker constant.

The dramatic decrease of the van der Waals attraction energy with the distance of particles (regardless of their shapes, see Parsegian 2006) is necessary to project to the values of the thermal energy $k_B T$, where k_B represents the Boltzmann constant (1.380649×10^{-23} J·K^{-1}) and T is the absolute temperature. For a classical temperature 20 °C ($T = 293.15$ K) the thermal energy is about 4.05×10^{-21} J. Comparing this value with the common values of the Hamaker constant $10^{-19} \div 10^{-20}$ J and the values in Fig. 1.6 we can see that a contribution of the van der Waals forces fades with higher distances between particles not only as such but also in comparison with other terms in the balance equation.

It is possible to derive a simplification of eq. (1.24). The approximate expressions

$$V_a^{appr}(s) = -0.038 \times (s - 1.9997)^{-1.15} \times A \equiv -0.038 \times (l_r + 0.0003)^{-1.15} \times A, \ l_r \in [0.001, 0.1] \tag{1.25}$$

$$V_a^{appr}(s) = -0.0089 \times (s - 1.85)^{-2.86} \times A \equiv -0.0089 \times (l_r + 0.15)^{-2.86} \times A, \ l_r \in [0.1, 1.1] \tag{1.26}$$

valid in two consecutive regions deviate almost negligibly from eq. (1.24). The deviation prevailingly does not exceed 4% or less; that is why a comparison illustrated in Fig. 1.7 is chosen only for selected points on the curve given by eq. (1.24) to present better legibility. Further, the functional behaviour is preserved, and hence a rate of decay of energy in dependence on l_r is apparent.

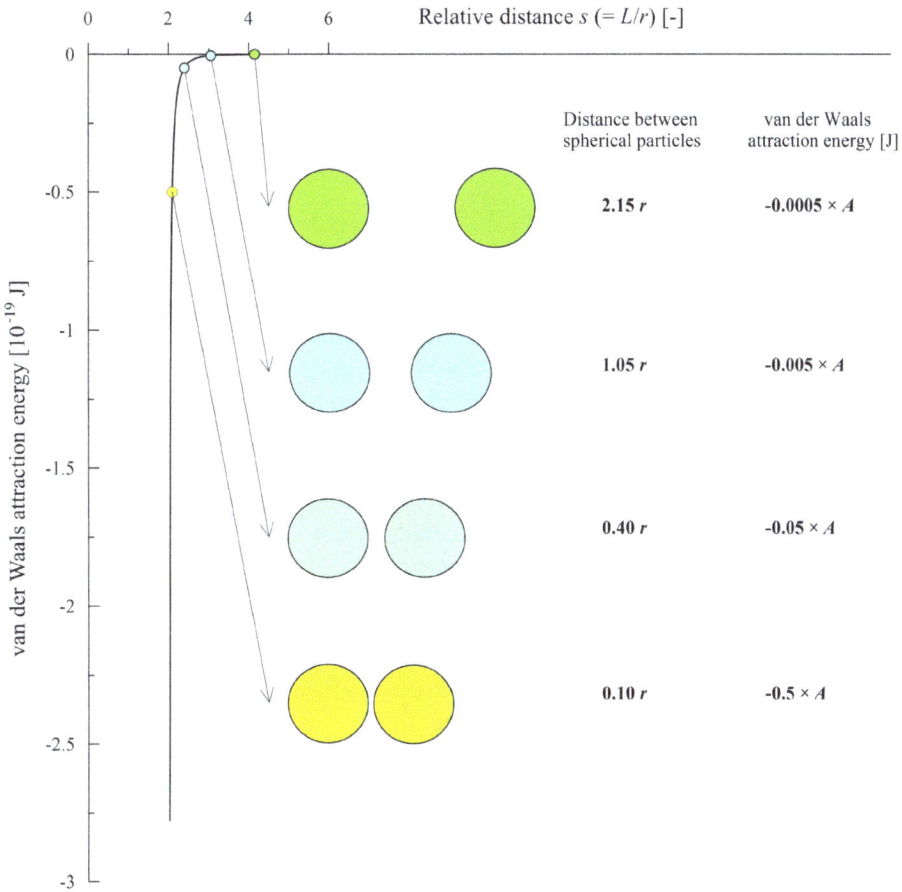

Fig. 1.6: A dramatic decay of van der Waals attraction energy with the increasing distance between equisized spherical particles.

The character of curves depicting the relative van der Waals attraction energy ($V_a(l_r,$ $q_r)/V_a(l_r,1)$) in dependence on the parameter q_r is strongly dependent on the distance between spherical particles. For all values l_r the curves exhibit a convex shape for low values of q_r (a combination of the smaller and larger particles) consecutively converting its shape to the concave one (see Fig. 1.8).

The travelling inflection points q_r^{infl} can be approximated (for $q_r \geq 0.05$) by the relation:

$$q_r^{infl} = 0.85 \times l_r^{0.57} \tag{1.27}$$

with the mean deviation 1.6% (see Fig. 1.9). The course of the individual relative curves can be approximated by eq. (1.28)

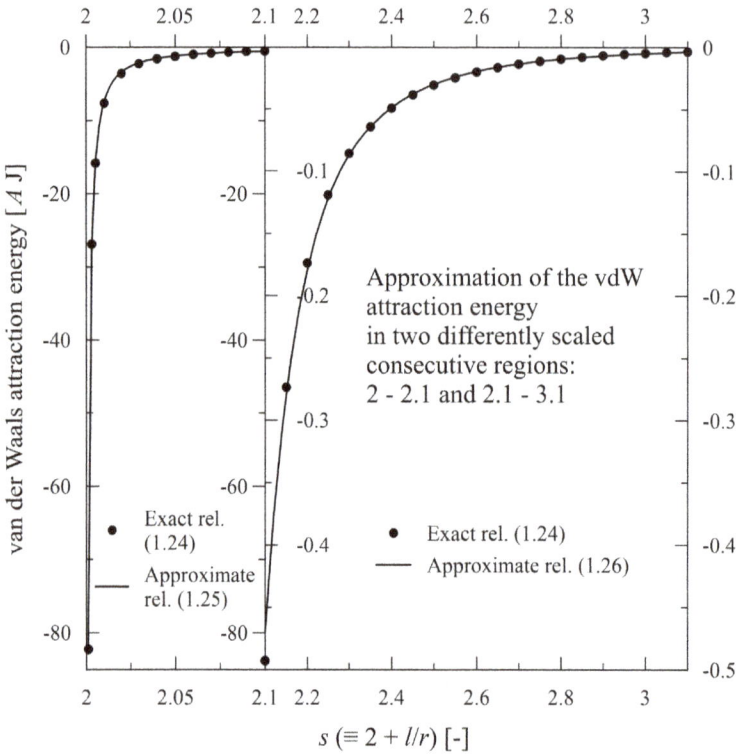

Fig. 1.7: The simplified approximate relations for the van der Waals attraction energy between two equisized spherical particles in dependence on their normalised distance $l_r = l/r$.

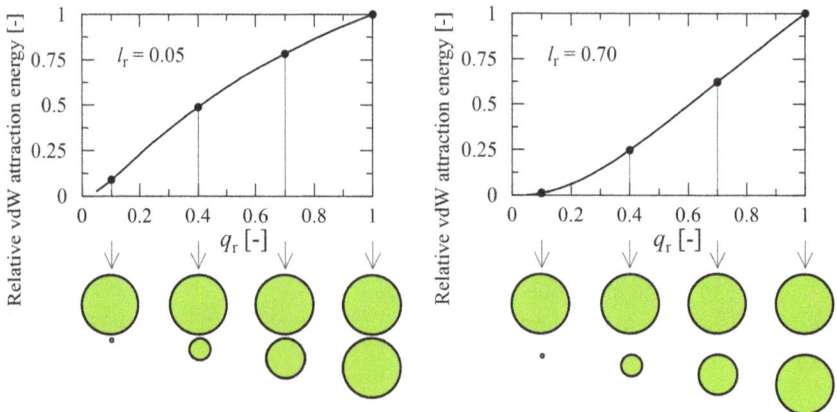

Fig. 1.8: A relative change of van der Waals attraction energy in dependence on the relative ratio of particles radii q_r and their relative distance l_r.

$$\text{Convex} - \text{concave curve} = \frac{V_a^{appr}(l_r, q_r)}{V_a^{appr}(l_r, 1)}. \tag{1.28}$$

The courses are invariant with respect to the values of the Hamaker constant A.

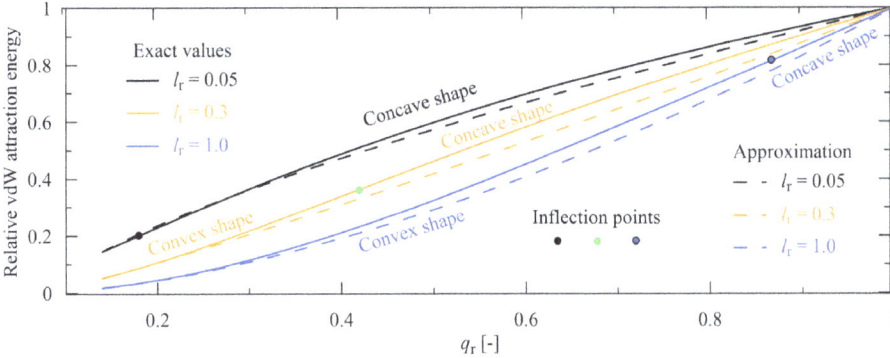

Fig. 1.9: A smooth passage of the relative van der Waals attraction energy from the dominant concave shape (close proximity of both spherical particles) to the dominant convex one (relatively more distant particles).

The quantum nature of light causes weakening of correlations in fluctuations of the charge and current densities in the objects or fluctuations of the fields (Boström et al. 2012). At sufficiently short distances, where the differences caused by the finite speed of light can be fully neglected, we speak about the discussed van der Waals force. However, at large distances – due to the finite speed of light (in spite of its value 299,800 km.s^{-1}) evoking the retardation effects – the so-called Casimir force is present (Casimir and Polder 1948). The extent of the retardation effects causes inapplicability of the Hamaker's formulae for different geometries when the distance exceeds approximately 3 nm.

Generally, there are two ways how to describe these situations: a usage of the Lifshitz approach (Lifshitz 1956) implicitly calculating with the retardation effects or a modification of the Hamaker's formulae by the multiplicative terms approximating a deviation of the strict Hamaker's formulae from real force distribution (respecting the retardation effects). The second alternative is based on two factors: a possibility of an experimental setting of the Hamaker constant and its knowledge for a variety of materials. As the interaction energy monotonically decreases with an increasing distance, it gives a hint what concerns the functional structure of the multiplicative terms. The various proposals are summarised in Gregory (1981).

1.4.2 A determination of the Hamaker constant

The above-mentioned analysis supposes the interaction of objects of the same material in a vacuum. If we suppose two different materials, then the Hamaker constant A_{12} is given by (Gregory 2006, p. 71)

$$A_{12} = \frac{27}{32} \cdot \frac{h\, v_1 v_2}{(v_1 + v_2)} \cdot \frac{n_1^2 - 1}{n_1^2 + 2} \cdot \frac{n_2^2 - 1}{n_2^2 + 2}, \tag{1.29}$$

where h is the Planck constant ($= 6.62607015 \times 10^{-34}$ [J·Hz$^{-1} \equiv$ kg·m^2·s^{-1}]) relating photon's energy with its frequency, v_1 and v_2 are the characteristic dispersion frequencies of the materials, and n_1 and n_2 are the refractive indices.

If both materials exhibit similar characteristics ($v_1 \sim v_2$, $n_1 \sim n_2$), rel. (1.29) can be simplified to the form:

$$A_{11} = \frac{27}{64} \cdot h\, v_1 \cdot \left(\frac{n_1^2 - 1}{n_1^2 + 2}\right)^2. \tag{1.30}$$

If we know the individual Hamaker constants A_{11} and A_{22} of two different materials, then it is possible – as the first approximation – to evaluate the composite Hamaker constant A_{12} as a geometrical mean

$$A_{12} = \sqrt{A_{11} \cdot A_{22}}. \tag{1.31}$$

However, this is still derived under the assumptions that both objects (particles) are placed in a vacuum.

To modify the whole process to the practical conditions, it is necessary to take into account the properties of the carrier liquids (as water in water treatment processes). Let us suppose that the Hamaker constants for two different media (particles 1 and 2) are A_{11} and A_{22} (see Fig. 1.10) and for the intervening medium (e.g. water) A_{33}.

Balancing the passage from the upper arrangement in the top part of Fig. 1.10 to the one in the bottom part we can derive a value for the real composite Hamaker constant:

$$A_{132} = A_{12} + A_{33} - A_{13} - A_{23} \tag{1.32}$$

where the individual A_{ij} represent the values in a vacuum. If we substitute the geometrical means of A_{ij} (rel. (1.31)) into rel. (1.32)

$$A_{132} = \sqrt{A_{11} \cdot A_{22}} + \sqrt{A_{33} \cdot A_{33}} - \sqrt{A_{11} \cdot A_{33}} - \sqrt{A_{22} \cdot A_{33}}, \tag{1.33}$$

we get

$$A_{132} = \left(\sqrt{A_{11}} - \sqrt{A_{33}}\right) \times \left(\sqrt{A_{22}} - \sqrt{A_{33}}\right), \tag{1.34}$$

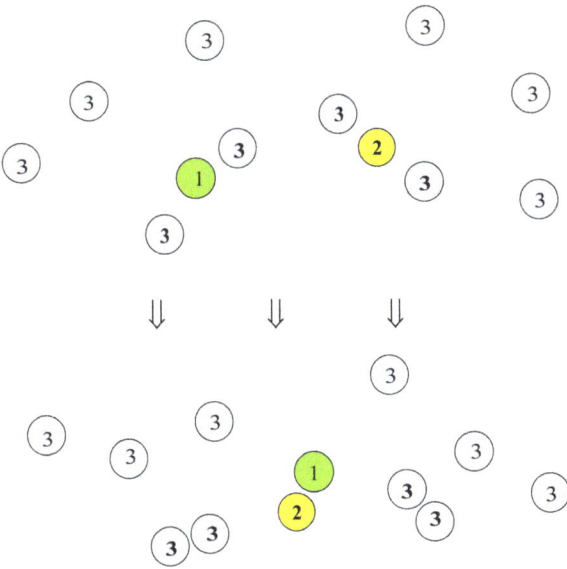

Fig. 1.10: Interaction of two different particles 1 and 2 in a medium 3.

and for the similar materials 1 and 2 we obtain the simplified relation

$$A_{132} = \left(\sqrt{A_{11}} - \sqrt{A_{33}} \right)^2. \tag{1.35}$$

Relation (1.34) approximately indicates the positive or negative sign (attraction or repulsion) of the van der Waals forces. For the carrier liquids with the lower Hamaker constant (as, for instance, water) in comparison with other materials, the van der Waals contribution is dominantly attractive. However, if the distance between two molecules is gradually reduced, at some distance (called the van der Waals radius), the attractive vdW forces are transforming to the repulsive forces.

In the preceding section, the determination of the Hamaker constant is carried out under the assumption of pairwise additivity of the individual intermolecular forces regardless the presence of the neighbouring molecules. This assumption is acceptable in gases; however, in the condensed media (liquids, solids) is no longer tenable. This assumption was eliminated by Lifschitz (1956) applying his macroscopic theory which ignores the atomic structure and works with the forces between large bodies (continuous media 1, 2, and 3) characterised by their dielectric constants (ε_1, ε_2, and ε_3) and refractive indices. In other words, microscopic composition of a material is characterised by its bulk dielectric response. As follows from the Lifschitz procedure, relatively small changes in dielectric behaviour can result in progressive changes in strength of the van der Waals interactions. It requires a responsible approach in the inevitable process of simplification of the resulting formulas.

The non-retarded Hamaker constant A (corresponding to the above-mentioned composite A_{132} supposing pairwise additivity) can be derived in a form of an infinite series where the higher-order terms (≥ 2) can be neglected (their contribution is less than 5%). The first terms representing the zero-frequency energy of the van der Waals interaction including the Keesom and Debye contributions and the dispersion energy including the London contribution are of the form (Israelachvili 2011, p. 258)

$$A = \frac{3}{4}kT\left(\frac{\varepsilon_1 - \varepsilon_3}{\varepsilon_1 + \varepsilon_3}\right) \cdot \left(\frac{\varepsilon_2 - \varepsilon_3}{\varepsilon_2 + \varepsilon_3}\right) + \frac{3h}{4\pi}\int\limits_{v_1}^{\infty}\left(\frac{\varepsilon_1(iv) - \varepsilon_3(iv)}{\varepsilon_1(iv) + \varepsilon_3(iv)} \cdot \frac{\varepsilon_2(iv) - \varepsilon_3(iv)}{\varepsilon_2(iv) + \varepsilon_3(iv)}\right) dv, \quad (1.36)$$

where $\varepsilon_m(iv)$, $m = 1,2,3$ are the values at imaginary frequencies and $v_1 = 2\pi kT/h$ (=3.71 × 10^{13} [s^{-1}] at 10 °C and 3.84 × 10^{13} [s^{-1}] at 20 °C).

Relation (1.36) can be simplified by expressing the dielectric permittivity by means of the frequencies. Further, supposing identity of continuous media 1 and 2 it is possible to obtain the simplified version of rel. (1.36) in the form (Israelachvili 2011, p. 261)

$$A = \frac{3}{4}kT\left(\frac{\varepsilon_1 - \varepsilon_3}{\varepsilon_1 + \varepsilon_3}\right)^2 + \frac{3hv_e}{16\sqrt{2}} \cdot \frac{\left(n_1^2 - n_3^2\right)^2}{\left(n_1^2 + n_3^2\right)^{3/2}}, \quad (1.37)$$

where v_e is the main electronic absorption energy in the UV typically around 3×10^{15} [s^{-1}].

A development of the instrumental techniques provides more accurate devices for an experimental determination of the van der Waals forces (Keesom, Debye, and London) and consequently the Hamaker constants. These values can be confronted with the purely theoretical results or their 'engineering' approximations. This way it can be also tested adequacy of the individual simplifications.

A historical review of surface force measurement techniques back to nearly the end of the twentieth century is presented in Craig (1997). However, in spite of efficient techniques such as atomic force microscopy (Butt et al. 2005), total internal reflection microscopy (Prieve 1999), optical laser tweezers (Lee et al. 2023) some questions for even otherwise common materials such as water seem to be still open (Gudarzi and Aboutalebi 2021). In this context, it is necessary to mention a difference between measurement of intervened gases and more challenging condensed materials (liquids, solids).

1.4.3 Rate of the attractive, drag, and gravitational forces

The significance of colloid interactions (including the van der Waals attraction energy) is documented in Fig. 1.11 using an example of a spherical particle in a close vicinity of the flat plane. Three forces – attractive, drag, and gravitational – increase with an increase of particle diameter d with the first, second, and third power of d, respectively. If in the initial case the mutual ratio of the inputs of these forces is 9:3:1 (a dominating contribution of the attractive force), then – if the diameter is tripled – the mutual ratio is equalised ($9 \times 3, 3 \times 3^2, 1 \times 3^3$) and purely colloid interactions cannot describe the behaviour of the particles. Consequently, it is necessary to have in mind that generally declared borderline 1 µm between colloidal and suspension structures regardless the material is rather a tentative convention than a serious turning point (see the differences between clay (1 µm) and algae (10 µm) materials caused by different densities given earlier). The densities regulate the corresponding initial cases and hence the real boundaries between the colloidal and suspension structures.

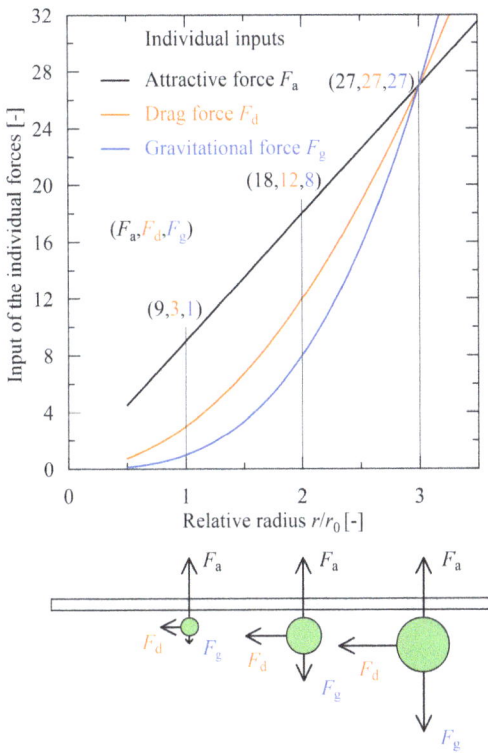

Fig. 1.11: A change of mutual contributions of the attractive, drag, and gravitational forces with an increase in particle diameter (equidistantly located along a flat plate). The triples indicate the changing weights of the individual inputs. The force arrows express mutually related magnitudes and not the absolute values (each magnitude increases with an increasing diameter).

1.5 Lennard-Jones potential

The method of the Lennard-Jones potential represents a classical intuitive approach to model the behaviour of two (and just two) interacting particles (atoms or small molecules), hence the term 'pair potential'. Intermolecular interactions are described by two inputs: attraction and repulsion. In the preceding section, it was introduced that each component of the attractive van der Waals forces (Keesom, Debye, and London) decreases with the sixth power of particles distance d, i.e., as d^{-6}. On the other hand, a decay of repulsive interaction energy can be supposed between d^{-9} and d^{-15} (Hiemenz and Rajagopalan 1997, p. 469). Based only on smoother calculating procedure, the exponent −12 is often used which leads to a notation 12–6 potential.

This way Lennard-Jones (Jones 1924a,b) expressed the total potential energy as

$$V_{LJ}(d) = A\,d^{-12} - B\,d^{-6}, \tag{1.38}$$

where $A > 0$ and $B > 0$ by convention, a parameter B cumulates the constants from the Keesom, Debye, and London expressions. Originally, the values −13.5 and −3 were used for the exponents. Later, after London's results concerning (nowadays denoted) the London forces and their decay (d^{-6}), Lennard-Jones changed the values of the exponents to the present 12–6 form. More details on the history of the Lennard-Jones potential are presented in Lenhard et al. (2024). The mutual comparison of both (attractive and repulsive) inputs in rel. (1.38) is schematically sketched in Fig. 1.12.

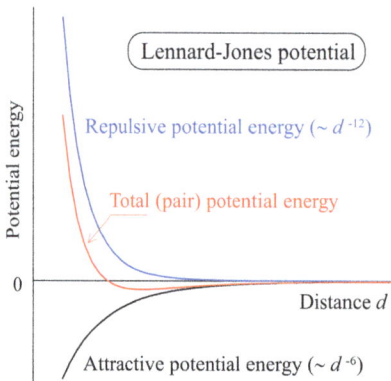

Fig. 1.12: Alternating dominancy of the repulsive and attractive components in the Lennard-Jones potential.

The repulsive potential energy dominates at very close vicinity (very low values of a distance d), this disbalance with the attractive potential energy is gradually levelled to a point where the total potential energy nullifies and consequently attains its minimum (an equilibrium state). Then, with higher mutual distances, both inputs tend to zero and a pair of particles is not mutually affected. These individual sketches can be converted to a single universal profile by introducing two new parameters ε and σ related to the parameters A and B through the expressions

$$\varepsilon = \frac{B^2}{4A}, \quad \sigma = \left(\frac{A}{B}\right)^{\frac{1}{6}}. \tag{1.39}$$

Based on these expressions rel. (1.38) is transformed to the form

$$V_{LJ}(d)/\varepsilon = 4 \times \left[\left(\frac{\sigma}{d}\right)^{12} - \left(\frac{\sigma}{d}\right)^{6}\right] \tag{1.40}$$

with clear physical meanings of both parameters. This universal profile is depicted in Fig. 1.13, where the total Lennard-Jones potential nullifies for $d = \sigma$ and $-\varepsilon$ represents a value of the minimal potential (unified to −1 in the universal graph). By differentiating the right-hand side of rel. (1.40) and equating to zero we obtain a location of the minimal potential $d/\sigma = 2^{1/6}$.

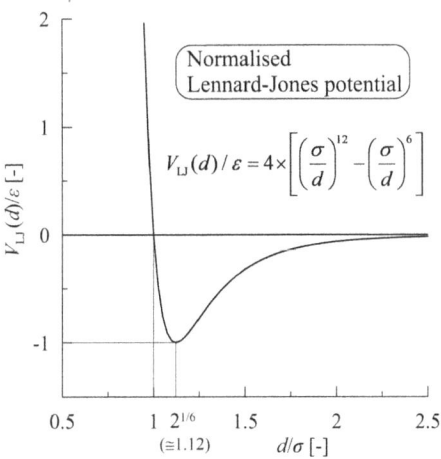

Fig. 1.13: A universal profile of the Lennard-Jones potential.

For return to the individual graphs (Fig. 1.12), it is necessary to use the inverse expressions

$$A = 4\varepsilon\sigma^{12}, \quad B = 4\varepsilon\sigma^{6}. \tag{1.41}$$

The preceding procedure documents a suitability of introducing 12 as a 'repulsive' exponent from the calculating viewpoint. However, the other values should be also considered as mentioned above. This point was addressed by Mie (1903) already in 1903 who introduced the general exponents n_{rel} and n_{rep} in eq. (1.38) which was further elaborated (for solids) by Grüneisen (1912). Other potential functions are summarised in Israelachvili (2011, Table 7.2).

1.6 Electrical double layer

Helmholtz (1853) was the first who realised that a charged electrode in the electrolyte solution repels the co-ions of the same charge from its surface and simultaneously attracts the counterions of the opposite charge. This way two layers of opposite polarity create an electrode-electrolyte interface.

A dependence of capacitance on the applied potential and the ionic concentration directed Gouy (1910) and Chapman (1913) to an introduction of the so-called diffuse layer not so compact as in the Helmholtz case. This enabled to describe an exponential decrease of potential with the distance from the interface. Interactions of ions (considered as point charges which polarisability is neglected) are supposed to be of purely electrostatic nature (no specific adsorption).

A problem with highly charged double layers inspired Stern (1924) to a combination of the two preceding approaches: a fixed (nowadays called) Stern layer of counterions adsorbed to the interface followed by a diffuse layer. The Stern layer was modified by Grahame (1947) who distinguished between two types of counterions in the closest vicinity of the interface. The first group is formed by the counterions without solvation shells with a direct contact with ions ('specifically adsorbed counterions') with the so-called inner Helmholtz plane passing the counterions centres. The second group is formed by the counterions with solvation shells separating them from the direct contact with the ions but still with a very strong bound to the interface. The plane connecting the centres of these counterions is denoted as the outer Helmholtz one. This second layer then converts to the Gouy-Chapman diffuse layer gradually balancing the ions and counterions presence to a level corresponding to the bulk region (bulk solution) (see Fig. 1.14 with the explanation in the figure caption). Hence, the notion electrical double layer (EDL) covers the charged surface together with a diffuse layer. More details are introduced in the monographs by Hiemenz and Rajagopalan (1997) and Israelachvili (2011).

The assumption that potential ψ attains low values (in the sense that electrical energy is exceeded by thermal energy) enables to derive a simple equation (Hunter 2001, p. 320) for a determination of ψ (so-called Debye-Hückel approximation). In this (Helmholtz-type) equation the Laplacian of ψ is related with ψ through a square of the introduced parameter κ (the Debye-Hückel parameter)

$$\Delta\psi = \kappa^2\psi, \tag{1.42}$$

where

$$\kappa = \left(\frac{e^2\Sigma n_i z_i^2}{\varepsilon k_B T}\right)^{\frac{1}{2}}, \tag{1.43}$$

and where e is the elementary charge (=$1.602176634 \times 10^{-19}$ C), n_i is the bulk concentration of ions of type i, z_i is the valency of ions of type i, ε is the permittivity of the solu-

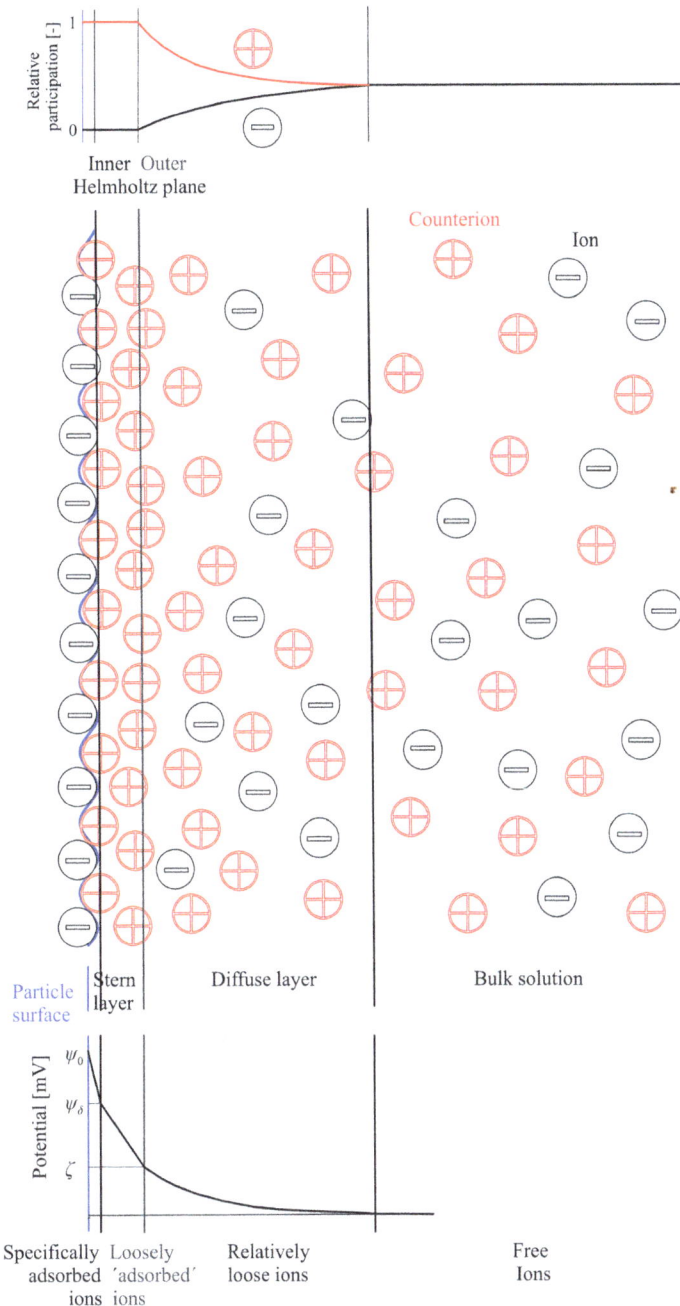

Fig. 1.14: Arrangement of the electrical double layer with the courses of potential and ion participation (anions vs. cations). The waviness of surface has to be replaced by a straight line in deriving the EDL characteristics because it is one of the principal assumptions. The question of surface unevenness will be discussed later; here it serves just for emphasising of the a priori assumptions (a perfectly smooth surface).

tion (relative permittivity × permittivity of a vacuum 8.8541878188(14) × 10^{-12} F·m^{-1}), k_B is the Boltzmann constant (= 1.380649 × 10^{-23} J·K^{-1}), and T is the absolute temperature.

The Debye-Hückel parameter represents a key factor in characterising the EDL as the so-called Debye length $\lambda_D = 1/\kappa$ determines an extent of the EDL.

In the case of planar surface, eq. (1.42) simplifies to

$$\frac{d^2\psi(x)}{dx^2} = \kappa^2\psi(x), \tag{1.44}$$

where x denotes the distance from the flat surface. Simultaneously the potential ψ has to fulfil two natural conditions:

$$\psi(x=0) = \psi_0; \quad \lim_{x\to\infty} \psi(x) = 0 \tag{1.45}$$

postulating a finite value at (strictly) planar surface and gradual fading of ψ with a distance. Under these conditions the solution of eq. (1.44) is of the form

$$\psi(x) = \psi_0 \cdot \exp(-\kappa x) \tag{1.46}$$

demonstrating an exponential smooth decrease with x (Hiemenz and Rajagopalan 1997, p. 510). To normalise rel. (1.46) with respect to the Debye length, it can be simply rewritten to the form

$$\psi(x) = \psi_0 \cdot \exp(-x/\kappa^{-1}). \tag{1.47}$$

From here we can see that $\psi/\psi_0 \cong 0.37$ at the Debye length κ^{-1}.

In the case of spherical surface an analogous solution of eq. (1.42) and boundary conditions, rel. (1.45), can be derived (Hiemenz and Rajagopalan 1997, p. 511):

$$\psi(s) = \psi_0 \cdot \left(\frac{R}{R+s}\right) \cdot \exp\left(-\frac{s}{\kappa^{-1}}\right), \tag{1.48}$$

where s denotes the distance from the spherical surface.

The values of the Debye length strongly subject to a solution concentration (see Fig. 1.15). For aqueous electrolyte solutions, Israelachvili (2011, p. 274) introduces the approximate typical Debye lengths for 10^{-3} M and 10^{-1} M solutions as 10 nm and 1 nm, respectively.

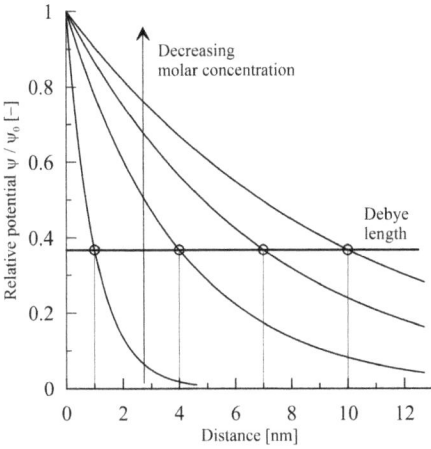

Fig. 1.15: An increasing Debye length with a decreasing molar mass.

1.7 The Derjaguin-Landau-Verwey-Overbeek (DLVO) theory

The Derjaguin, Landau, Verwey, and Overbeek (DLVO) theory (Derjaguin and Landau 1941; Verwey and Overbeek 1948) describes particle interactions combining two long-range inputs: the attractive van der Waals forces and the repulsive forces generated by overlapping of EDLs (see Fig. 1.16). This approach is beneficial in many cases where these two inputs (or apolar interaction energies) dominate.

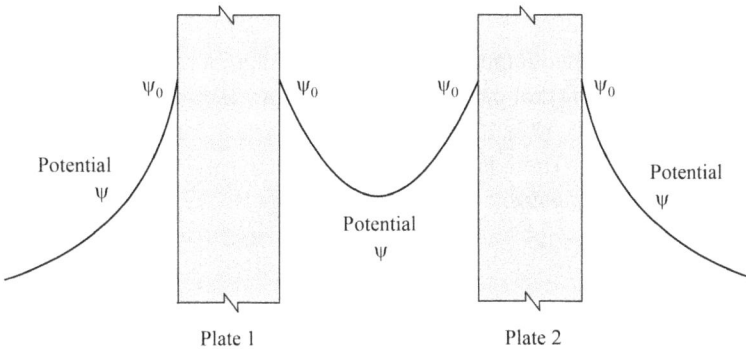

Fig. 1.16: A course of potential ψ between two plates created by overlapping of two electrical double layers.

The problem is that in many cases this classical DLVO approach rather fails. There are two reasons:

a) improper assumptions during derivation of the expressions describing these two forces such as a perfect smoothness of the studied objects up to a molecular level;

b) existence of other forces comparable (or even exceeding) with the individual DLVO inputs.

a) Surface roughness

Figure 1.14 intentionally emphasises one of the crucial assumptions in deriving both van der Waals forces and EDL description – no variation in surface topography. In other words, the theory of both interaction energies a priori supposes perfectly smooth and geometrically regular surfaces to a molecular level. It means that wavy character of surface should be replaced by a straight line (projection of a plane surface) resulting in the corresponding arrangement of the inner Helmholtz plane, which intersects the adsorbed layer of the counterions in their centres.

Surface roughness exhibits an apparent influence on both DLVO characteristics (Bhattacharjee et al. 1998; Walz 1998; Hoek et al. 2003; Thormann 2017). The values for rough surface are lower in both cases, particularly at very short separation distances. However, a decrease differs. While the van der Waals forces are always significantly reduced, the EDL force is mostly affected at high salt concentrations, when the magnitudes of the Debye length and amplitude roughness are comparable. It means that the size of the asperities and their distribution (densities) on the surface play a significant role. Among other things, the reduction in the energy barrier is strongly correlated with the magnitude of surface roughness.

Practically all surfaces (apart from mica) exhibit higher or lower degree of unevenness, and hence the derived relations for both interaction energies should be considered as the approximate ones with various deviations from the experimental measurements.

b) Existence of other forces

Usefulness of the DLVO approach to modelling colloids in water is rather limited due to the fact (as introduced above) that participation of (Keesom, Debye, London) forces in water is in the ratio (84.8, 4.5, 10.5). It means that only approximately 1/10 of forces is covered by the DLVO theory. There is a series of other short- and long-range effective forces that are at least comparable with both DLVO inputs. These forces are discussed below.

1.8 The eXtended DLVO (XDLVO) theory

After an initial success of the DLVO theory considering apolar interaction energies there appeared a necessity to extent this approach by other forces and not to be limited only to two classical inputs (van Oss 1994; Elimelech et al. 1995; Hiemenz and Rajagopalan 1997; Ninham 1999 (involves a detailed summary of the DLVO assumptions between planar surfaces), Ohki and Ohshima 1999; Boström et al. 2001; van Oss 2008; Israelachvili 2011; and references therein). Ninham (2019) even says: "Indeed if anyone claims agreement with DLVO theory, his measurements are wrong."

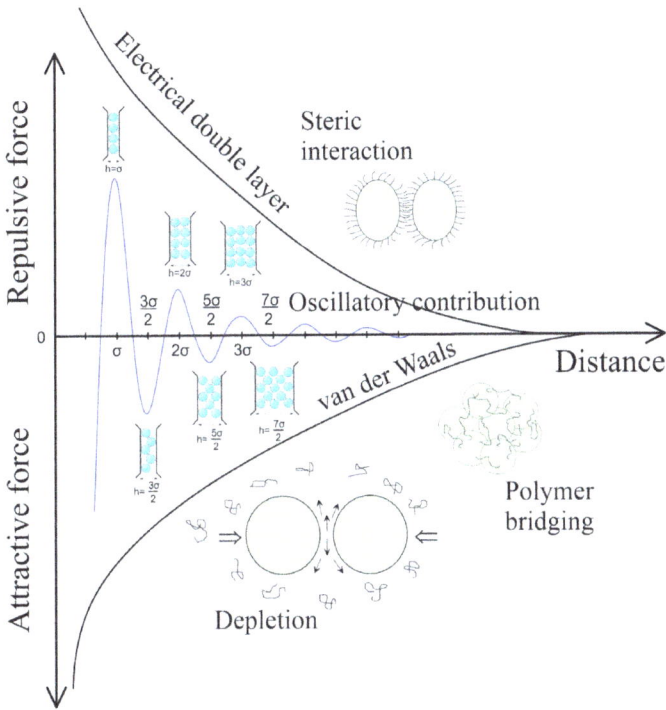

Fig. 1.17: Illustration of selected XDLVO forces.

A procedure considering other force(s) is called the eXtended DLVO (XDLVO) theory (the term coined by van Oss 1993) and the individual contributions are denoted as extended DLVO (XDLVO) or non-DLVO interactions. In the following the selected XDLVO interactions are described in more detail, and some of them are illustrated in Fig. 1.17.

1.8.1 Lewis acid-base interaction

Lewis (1916) formulated the chemical bond with a pair of shared electrons and introduced – at present denoted as – the Lewis base as a substance possessing a lone pair of electrons which may be used to complete the stable group of another atom, the Lewis acid as a substance which can employ a lone pair from another molecule in completing the stable group of one of its own atoms. Lewis' formulation was in a compliance with the 'discrete' planetary model of an atom at that time (point particles moving along classical trajectories). Hence, acids are electron-pair acceptors and bases are electron-pair donors.

In a language of quantum physics the formation of an electron pair should be replaced by the lowering of the kinetic energy density of the shared electrons in the bonding region, which is provided by the interference of the atomic wavefunctions (matter fields continuously extended over space). The whole process, expressed using the molecular orbitals, is depicted in Fig. 1.18. The Lewis acid (Lewis 1923, 1938) is characterised by an empty frontier orbital exhibiting the lowest energy which is called the lowest unoccupied molecular orbitals (LUMO). On the other hand, the Lewis base has the frontier orbital which is filled and exhibits the highest energy and is called the highest occupied molecular orbital (HOMO). LUMO is capable of accepting the electron pair from HOMO to form a Lewis adduct (a direct addition of the molecules containing all atoms of the single molecules). Hence, the definitions of Lewis acid and base can be reformulated as follows:

– A Lewis base is any substance which has electron density that can be shared with another substance in a chemical reaction.
– A Lewis acid is any substance capable of accepting electron density from a Lewis base.

Therefore, a covalent bond is created when the electron density of the electron pair localised on the base flows into a molecular orbital in the acid and is thus shared by both species (Jensen 2016). The two orbitals are mixed if they at least overlap each other, and they have the proper symmetry and approximately the same energy (Jensen 2016).

It has been proved (Brant and Childress 2002; Lin et al. 2014; Huang et al. 2021a,b; Wang et al. 2021; Liu et al. 2022) that the Lewis short-range acid-base (AB) (electron donor/electron acceptor) polar interaction represents an optimal interaction which should be implemented to the two classical ones (vdW and EDL). Importance of this interaction documents a mutual participation of the AB and vdW terms in a descrip-

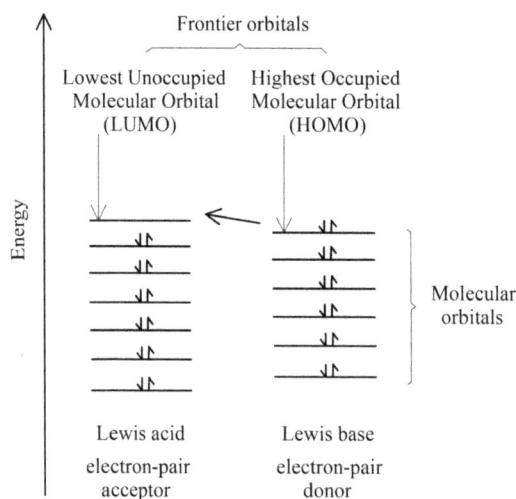

Fig. 1.18: Lewis process of an origin of the covalent bond: acids are electron-pair acceptors and bases are electron-pair donors.

tion of the total surface tension in the case of water (polar liquid for which Keesom forces dominate the London ones). The AB contribution (51 mJ/m^2 out of the total surface tension 72.8 mJ/m^2) more than doubles the vdW one (in the region not exceeding the units of nm). Hence, this combination of the XDLVO approach (AB, vdW, EDL) results, for example, in more accurate explanation of the bacterial adhesion.

Detailed information on Lewis A-B interaction can be found in Laurence and Gal (2010), Jensen (2016), and Zhao et al. (2019).

1.8.2 Born repulsion

Asymptotic behaviour of the attractive van der Waals forces in approaching a surface is compensated by very short-range repulsive Born forces. The forces originate by overlapping molecular orbitals and completely vanish at a distance 0.5–1 nm (see Fig. 1.19). It means that their field of activity more or less does not exceed molecular dimensions. For comparison a diameter of water molecule attains approximately 0.275 nm.

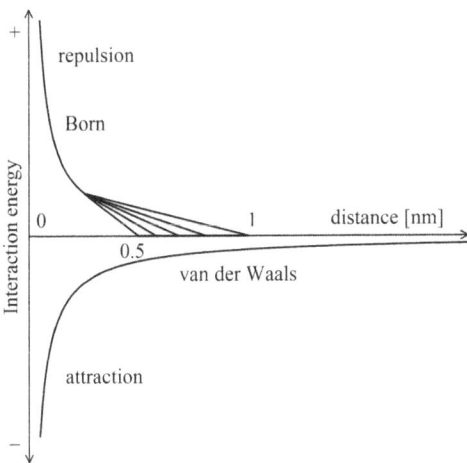

Fig. 1.19: Interplay between the Born and van der Waals forces with an indication of the vanishing zone.

Among other things, mutual interaction between the Born and van der Waals forces at a close vicinity of a surface prevents the particles against penetrating into a surface.

1.8.3 Solvation

According to the IUPAC terminology (IUPAC 2006) solvation is characterised as: "Any stabilizing interaction of a solute (or solute moiety) and the solvent or a similar interaction of solvent with groups of an insoluble material (i.e. the ionic groups of an ion-exchange resin). Such interactions generally involve electrostatic forces and van der Waals forces, as well as chemically more specific effects such as hydrogen bond formation."

Solvation is a kinetic process quantified by its rate, usually in mol/s, in contrast to solubility quantifying the dynamic equilibrium state (equating the rates of dissolution and precipitation, usually in mg/mL). The word 'solvation' is replaced by the term 'hydration' in the case when a solvent is water.

Solvation forces are not governed only by a relation solvent vs. solute but are also strongly influenced by the chemical and physical properties of the wall surfaces. The mere DLVO description is usually no longer valid if surfaces or particles get closer (Israelachvili 2011, p. 341); Christenson (1988) introduces 5 nm. In this case, other, non-DLVO, interactions also enter the balance relations including solvation (hydration) forces with their magnitudes equalling or exceeding the classical DLVO ones.

The constraining effect of two solid surfaces exhibits a strong impact on the liquid density distribution function which is further reflected in the solvation pressure distribution. This can be illustrated by means of the most general type of solvation force: the oscillatory force, which is generated by an arrangement of solvent molecules. This force has its influence when a distance between surfaces attains a limited number of solvent molecules, with a higher number gradually fades. Correspondingly the magnitudes of the maxima and minima of the solvation (hydration) pressure decrease with increasing surface separation and the period of the oscillations is close to the average molecular diameter σ, at least beyond the first three layers of molecules (Christenson 1988) (see Fig. 1.20). Possible molecular arrangements in Fig. 1.20 are illustrative; more on this topic is presented in Israelachvili (2011, p. 346).

1.8.4 Hydrophobic attraction

The term 'hydrophobic' is generally connected with water repulsion. However, in reality hydrophobic surfaces attract water as explained in Hildebrand (1979). It can be documented on an example of Teflon pans. Such hydrophobic material – if immersed in water – attracts its molecules (a competition with the EDL) with a free energy of about 40–50 mJ/m^2. The hydrophobic effect denotes the attraction between two substances with nonpolar surfaces or with one nonpolar and one polar surface in water. The hydrophobic attraction represents the strongest non-covalent force between particles immersed in aqueous solution exceeding both van der Walls and EDL interactions.

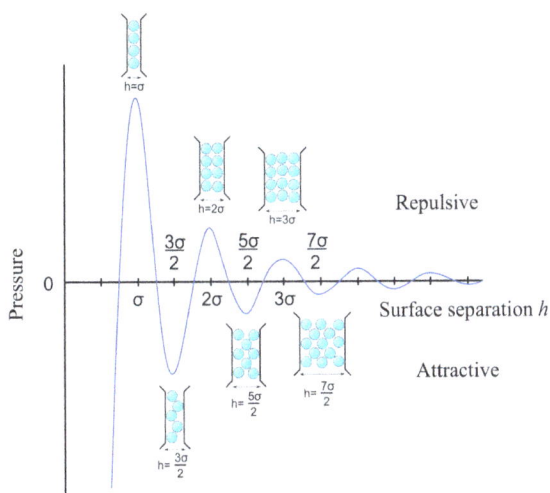

Fig. 1.20: Schematic dependence of pressure on a distance between planar surfaces.

After introducing the hydrophobic tendency to aggregate by Laskowski and Kitchener (1969) and Blake and Kitchener (1972) characterised by a positive change of free energy of the surrounding solvent (water), the ongoing development of this term has not been straightforward. The crucial point is in the experimental setting of this quantity. The hydrophobic force $F_{hydrophobic}$ is given by the total surface force $F_{surface}$ acting between two solids reduced by a sum of all other known forces F_{other}:

$$F_{hydrophobic} = F_{surface} - \Sigma F_{other}. \tag{1.49}$$

The natural problem is with identifying these other forces and especially with their quantifying (Ducker and Mastropietro 2016). This difficulty – among other things – also evokes an uncertainty about a possible range of hydrophobic interaction: whether short-range or long-range (taken with respect to a size of water molecule – about 0.275 nm). Ducker and Mastropietro (2016) after analysing the experimental works published before their contribution incline to the upper limit of 5 nm. This limit simultaneously with a comparison between the hydrophobic and van der Waals forces also determines possible applicability of the hydrophobic effect in practice.

Another way is described in Kékicheff (2019), where the hydrophobic surfaces studied experimentally are distributed into three groups: (1) inherently hydrophobic surfaces, (2) substrates coated with a hydrophobising agent bonded chemically onto the surface, (3) surfaces obtained with surfactant monolayers physically adsorbed from solution to deposited ones. In the case of group 2 it was shown that instead of a seemingly measured hydrophobic force was in reality measured a capillary force.

This was caused by the presence of the gaseous bridge formed from the coalescence of nanobubbles (nanosegments, nanocaps) at the liquid-hydrophobic interface.

The detailed analysis of the hydrophobic interactions is presented, e.g., in van Oss (2003) and Israelachvili (2011, Sect. 15.9) documenting still rather unclear origin of the hydrophobic force.

1.8.5 Steric interaction

Dispersion stability can be – relatively easy – improved by applying polymeric material as first presented by Fischer (1958). Nonionic polymer adsorbs on the surface of particles and creates a hull of thickness δ round the particle formed by dangling polymer chains. If two particles are approached closer than 2δ, the chains of both particles either mutually interpenetrate or are mutually compressed (see Fig. 1.21). In practice, the most frequent case is formed by a combination of both: interpenetration and compression. A key factor is solubility of the particles in a used solvent indicated by the polymer-solvent interaction parameter χ (introduced in the Flory-Huggins solution theory):

$$\chi = \frac{V_{\text{chains}}(\delta_s - \delta_p)^2}{RT},$$

(1.50)

where V_{chains} is the volume occupied by the polymer chains, δ_s and δ_p are the Hildebrand solubility parameters, R is the gas constant, and T is the temperature. If the solubility parameters are relatively close to each other in such a way that in a combination with other quantities in rel. (1.50) the resulting value of the parameter $\chi < 0.5$, then the particles are in a good solvent.

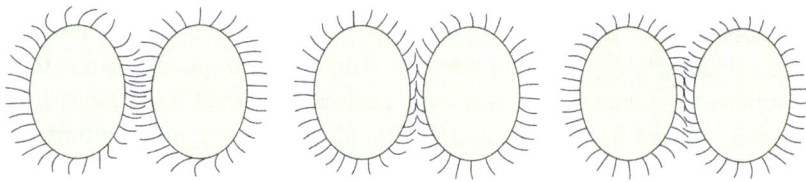

Fig. 1.21: Mutual location of dangling polymer chains in an overlapping position of two particles with adsorbed polymer: interpenetration (left), compression (middle), a combination of both (right).

Presence of a good solvent has an impact on the local increase of chains density in the interaction region. This results in a repulsion of both particles caused dominantly by the following factors:

- A good solvent enables smooth mixing of the polymer chains in the overlap region which causes an increase of the osmotic pressure; in this case a free energy of interaction G_{mix} characterising osmotic repulsion is positive.
- Simultaneously the configurational entropy of the chains in the interaction region is reduced due to a decrease in the volume available for interpenetrated or compressed polymer chains, and this volume restriction interaction is quantified by a free energy of interaction G_{el}.

The total steric interaction free energy G_{steric} can be approximated by two dominant contributions:

$$G_{steric} = G_{mix} + G_{el}. \tag{1.51}$$

While G_{el} always attains positive values, a sign of G_{mix} subjects to a mutual relation between the particles and a used solvent. For good solvents ($\chi < 0.5$) G_{mix} is always positive. However, for poor solvents ($\chi > 0.5$) G_{mix} attains negative values reflected in attractive mixing interaction. Nevertheless, even in this case there can be exhibited dispersion stability if positive G_{el} outweighs negative G_{mix} (Tadros 2010, p. 10).

Separating both particles by their polymer hulls has an unfavourable impact on the attractive van der Waals forces due to their attenuation with a distance. As these forces are proportional to the particles dimension, their sizes should be accompanied by a thicker (an increasing δ) polymer hull (Gregory 2006, p. 90).

The overall characterisation of steric interaction along with a particle-solvent relation and density of dangling polymer chains also depends on the coverage of polymer on particle's surface – whether adsorbed or grafted, see Israelachvili (2011, p. 387).

1.8.6 Polymer bridging

Steric interaction (particles repulsion) is achieved by relatively short polymers. This can be used in pharmaceutical, cosmetic, and food industry. On the other hand, for instance, in water treatment, mineral processing or paper manufacturing the opposite effect is required, i.e., a formation of particle clusters. To this aim longer polymers should be applied which contributes to so-called polymer bridging when adsorbing polymers have capability to bind more particles together and exceed repulsive interactions (see Fig. 1.22).

The process of polymer bridging depends – among other things – on the particle and polymer concentrations and their relative sizes. The topic of a stable-unstable-restabilisation transition with increasing polymer concentration is discussed in Gonzales García et al. (2020). Another parameter strongly participating in the strength of adsorption, interparticle forces, and cluster size is polymer molecular weight (Gallegos et al. 2023). For successful polymer bridging the molecular weights exceed-

Fig. 1.22: Adsorbing polymers connecting the individual particles, formation of particle clusters.

ing 10 millions are used. Such polymer macromolecules can exhibit more than 100,000 individual segments capable of adsorbing on the surface (Hogg 2013). In these cases, the possible reversible adsorption/desorption steps can be neglected as such polymers as a whole exhibit polymer bridging regardless of individual desorption. Clustering by polymer bridging can be also optimised by pH adjustment contributing to starting a coagulation process and forming small clusters, and hence, under re- duced concentration of polymer (a decreased number of necessary linkages among pre-clusters).

1.8.7 Depletion

Interplay between dispersed particles and added polymer material can be roughly di- vided into two main cases:
- Polymer segments can adsorb onto the particle surface (short polymer – steric repulsion → stabilisation, long polymers – polymer bridging → destabilisation).
- Polymer molecules do not adsorb onto the particle surface.

First contributions reporting the effective attractions between the colloidal particles caused by non-adsorbing macromolecules date back to the 1950s (Asakura and Oosawa 1954, 1958). The term 'depletion' used for this phenomenon was coined in Feigin and Napper (1980). Figure 1.23 depicts topology of this process. Sometimes the term depletion is equivalently substituted by a term 'negative adsorption'. If two non- adsorbing particles are at a sufficient distance from each other, then all polymer mol- ecules are in a bulk region with a possibility to 'travel' in between both particles. The polymer segment concentration can be approximated by a product of a Heaviside step function (attaining 0 in the interval (0,δ) and 1 for distances (from particle surface) exceeding δ) and a bulk polymer concentration. In a dilute polymer solution a depletion thickness roughly corresponds to the radius of gyration of the polymer chains. If we consider a single spherical particle, then an osmotic pressure is isotropic. This situation dramatically changes when two spherical particles mutually approach which results in an overlapping of the depletion layers. An isotropic character of the osmotic pressure is violated and a passage to an anisotropic condition results in a forming of a net osmotic force (indicated by the yellow arrows in Fig. 1.23). The polymer molecules are expelled to the bulk region jointly with a solvent for which it is favourable from the energetical viewpoint to solvate the polymers in the bulk region. This way the depletion interaction is generated by maximising the entropy of a colloid-polymer dispersion (Feng et al. 2015) and contributes to a destabilisation of the whole system due to an attraction of both particles.

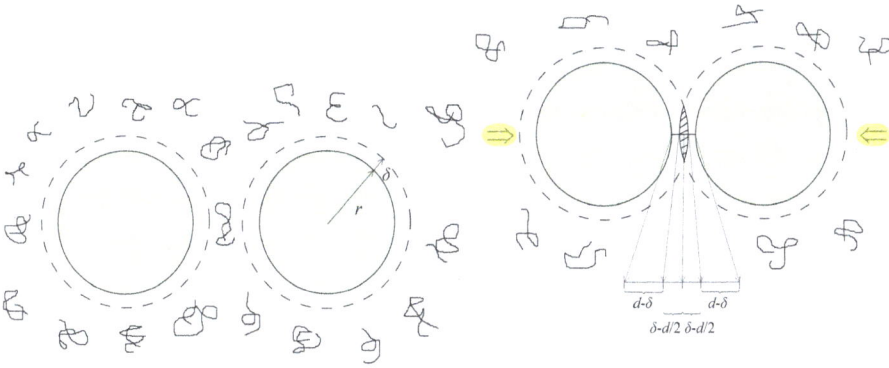

Fig. 1.23: Two couples of colloidal spheres with non-adsorbing polymers with the depletion spherical hulls of the thickness δ. Isotropic distribution of the osmotic pressure is gradually unbalanced when the spherical caps of the depletion hulls start to penetrate mutually. The yellow arrows depict the excess pressure.

The effective free energy of interaction is given by a product of the osmotic pressure and the overlapped volume indicated by the hatched volume formed by two spherical caps with a corresponding radius $r + \delta$ and a height $h = \delta - d/2$, where d denotes the distance between the particles' surfaces. From the general formula for a volume of the spherical cap

$$V_{cap} = \frac{\pi h^2}{3} \times [3(r + \delta) - h] \tag{1.52}$$

we obtain a relation for the hatched volume

$$V = \frac{\pi(2\delta - d)^2}{12} \times (6r + 4\delta + d).\tag{1.53}$$

The historical evolution of the Asakura-Oosawa theory is described in Oosawa (2021) and Miyazaki et al. (2022). The detailed analysis is presented in the monograph by Lekkerkerker et al. (2024, open access) which represents the substantially enlarged version of the first edition (Lekkerkerker and Tuinier 2011).

References

Anusree, K.V., Latha, G.M. (2023) Characterization of sand particle morphology: state-of-the-art. Bulletin of Engineering Geology and the Environment 82, 269. https://doi.org/10.1007/s10064-023-03309-x

Asakura, S., Oosawa, F. (1954) On interaction between two bodies immersed in a solution of macromolecules. Journal of Chemical Physics 22, 1255–1256. https://doi.org/10.1063/1.1740347

Asakura, S., Oosawa, F. (1958) Interaction between particles suspended in solutions of macromolecules. Journal of Polymer Science 33, 183–192. https://doi.org/10.1002/pol.1958.1203312618

Bargeman, D., van Voorst Vader, F. (1972) Van der Waals forces between immersed particles. Journal of Electroanalytical Chemistry and Interfacial Electrochemistry 37, 45–52. https://doi.org/10.1016/S0022-0728(72)80213-4

Bhattacharjee, S., Ko, C.-H., Elimelech, M. (1998) DLVO interaction between rough surfaces, Langmuir 14, 3365–3375. https://doi.org/10.1021/la971360b

Blake, T.D., Kitchener, J.A. (1972) Stability of aqueous films on hydrophobic methylated silica. Journal of the Chemical Society, Faraday Transactions 1: Physical Chemistry in Condensed Phases 68, 1435–1442. https://doi.org/10.1039/F19726801435

Blott, S.J., Pye, K. (2008) Particle shape: a review and new methods of characterization and classification. Sedimentology 55, 31–63. https://doi.org/10.1111/j.1365-3091.2007.00892.x

Boström, M., Sernelius, B.E., Brevik, I., Ninham, B.W. (2012) Retardation turns the van der Waals attraction into a Casimir repulsion as close as 3 nm. Physical Review A, Atomic, Molecular, and Optical Physics 85, 010701. https://doi.org/10.1103/PhysRevA.85.010701

Boström, M., Williams, D.R.M., Ninham, B.W. (2001) Specific ion effects: Why DLVO theory fails for biology and colloid systems. Physical Review Letters 87, 168103. https://doi.org/10.1103/PhysRevLett.87.168103

Bowen, W.R., Jenner, F. (1995) The calculation of dispersion forces for engineering applications. Advances in Colloid and Interface Science 56, 201–243. https://doi.org/10.1016/0001-8686(94)00233-3

Brant, J.A., Childress, A.E. (2002) Assessing short-range membrane–colloid interactions using surface energetics. Journal of Membrane Science 203, 257–273. https://doi.org/10.1016/S0376-7388(02)00014-5

Butt, H.-J., Cappella, B., Kappl, M. (2005) Force measurements with the atomic force microscope: Technique, interpretation and applications. Surface Science Reports 59, 1–152. https://doi.org/10.1016/j.surfrep.2005.08.003

Casimir, H.B.G., Polder, D. (1948) The influence of retardation on the London–van der Waals forces. Physical Review 73, 360–372. https://doi.org/10.1103/PhysRev.73.360

Chapman, D.L. (1913) A contribution to the theory of electrocapillarity. The London, Edinburgh, and Dublin Philosophical Magazine and Journal of Science 25, 475–481. https://doi.org/10.1080/14786440408634187

Christenson H.K. (1988) Non-DLVO forces between surfaces – solvation, hydration and capillary effects. Journal of Dispersion Science and Technology 9, 171–206. https://doi.org/10.1080/01932698808943983

Corey, A.T. (1949) Influence of Shape on Fall Velocity of Sand Grains. MSc Thesis, Colorado A&M College, 102 pp. https://hdl.handle.net/10217/195976

Craig, V.S.J. (1997) An historical review of surface force measurement techniques. Colloids and Surfaces A: Physicochemical and Engineering Aspects 129–130, 75–93. https://doi.org/10.1016/S0927-7757(97)00029-0

Debye, P. (1920) Die van der Waals'schen Kohasionskräfte. Nachrichten von der Gesellschaft der Wissenschaften zu Göttingen, Mathematisch-physikalische Klasse 55–73. https://gdz.sub.uni-goettingen.de/id/PPN252457811_1920?tify=%7B%22pages%22%3A%5B61%5D%2C%22view%22%3A%22export%22%7D

Derjaguin, B., Landau, L.D. (1941) Theory of the stability of strongly charged lyophobic sols and of the adhesion of strongly charged particles in solutions of electrolytes. Acta Physicochimica U.R.S.S. 14, 633–662. https://doi.org/10.1016/0079-6816(93)90013-L

Ducker, W.A., Mastropietro, D. (2016) Forces between extended hydrophobic solids: Is there a long-range hydrophobic force? Current Opinion in Colloid & Interface Science 22, 51–58. https://doi.org/10.1016/j.cocis.2016.02.006

Elimelech, M., Gregory, J., Jia, X., Williams, R.A. (1995) Particle Deposition and Aggregation. Measurement, Modelling and Simulation. Butterworth-Heinemann, Woburn, Massachusetts, USA.

Feigin, R.I., Napper, D.H. (1980) Stabilization of colloids by free polymer. Journal of Colloid Interface Science 74, 567–571. https://doi.org/10.1016/0021-9797(80)90226-X

Feng, L., Laderman, B., Sacanna, S., Chaikin, P. (2015) Re-entrant solidification in polymer-colloid mixtures as a consequence of competing entropic and enthalpic attractions. Nature Materials 14, 61–65. https://doi.org/10.1038/nmat4109

Fischer, E.W. (1958) Elektronenmikroskopische Untersuchungen zur Stabilität von Suspensionen in makromolekularen Loesungen. Kolloid-Zeitschrift 160, 120–141. https://doi.org/10.1007/BF01503288

Gallegos, M.J., Soetrisno, D.D., Samghabadi, F.S., Conrad, J.C. (2023) Effects of polymer molecular weight on structure and dynamics of colloid–polymer bridging systems. Journal of Physical Chemistry B 127, 3969–3978. https://doi.org/10.1021/acs.jpcb.3c01135

Gonzales García, A., Nagelkerke, M.M.B., Tuinier, R., Vis, M. (2020) Polymer-mediated colloidal stability: on the transition between adsorption and depletion. Advances in Colloid and Interface Science 275, 102077. https://doi.org/10.1016/j.cis.2019.102077

Gouy, G (1910) Sur la constitution de la charge électrique à la surface d'un électrolyte. Journal de Physique Théorique et Appliquée 9, 457–468. https://doi.org/10.1051/jphystap:019100090045700

Grahame, D.C. (1947) The electrical double layer and the theory of electrocapillarity. Chemical Reviews 41, 441–501. https://doi.org/10.1021/cr60130a002

Gregory, J. (1981) Approximate expressions for retarded van der Waals interaction. Journal of Colloid and Interface Science 83, 138–145. https://doi.org/10.1016/0021-9797(81)90018-7

Gregory, J. (2006) Particles in Water. Properties and Processes. Taylor and Francis.

Grüneisen, E. (1912) Theorie des festen Zustandes einatomiger Elemente. Annalen der Physik 344, 257–306. https://doi.org/10.1002/andp.19123441202

Gudarzi, M.M., Aboutalebi, S.H. (2021) Self-consistent dielectric functions of materials: Toward accurate computation of Casimir–van der Waals forces. Science Advances 7, eabg2272. https://www.science.org/doi/10.1126/sciadv.abg2272

Hamaker, H.C. (1937) The London – van der Waals attraction between spherical particles. Physica 4, 1058–1072. https://doi.org/10.1016/S0031-8914(37)80203-7

Helmholtz, H. (1853) Ueber einige Gesetze der Vertheilung elektrischer Ströme in körperlichen Leitern mit Anwendung auf die thierisch-elektrischen Versuche. Annalen der Physik und Chemie 165, 211–233. https://doi.org/10.1002/andp.18531650603

Hiemenz, P.C., Rajagopalan, R. (1997) Principles of Colloid and Surface Chemistry. 3rd ed., CRC Press, Taylor & Francis Group, Boca Raton.

Hildebrand, J.H. (1979) Is there a "hydrophobic effect"? Proceedings of the National Academy of Sciences 76, 194–194. https://doi.org/10.1073/pnas.76.1.194

Hoek, E.M.V., Bhattacharjee, S., Elimelech, M. (2003) Effect of membrane surface roughness on colloid −membrane DLVO interactions. Langmuir 19, 4836–4847. https://doi.org/10.1021/la027083c

Hogg, R. (2013) Bridging flocculation by polymers. KONA Powder and Particle Journal No. 30. https://doi.org/10.14356/kona.2013005

Huang, Z., Liu, J., Liu, Y., Xu, Y., Li, R., Hong, H., Shen, L., Lin, H., Liao, B.-Q. (2021a) Enhanced permeability and antifouling performance of polyether sulfone (PES) membrane via elevating magnetic Ni@MXene nanoparticles to upper layer in phase inversion process. Journal of Membrane Science 623, 119080. https://doi.org/10.1016/j.memsci.2021.119080

Huang, Z., Zeng, Q., Liu, Y., Xu, Y., Li, R., Hong, H., Shen, L., Lin, H. (2021b) Facile synthesis of 2D TiO2@MXene composite membrane with enhanced separation and antifouling performance. Journal of Membrane Science 640, 119854. https://doi.org/10.1016/j.memsci.2021.119854

Hunter, R.J. (2001) Foundations of Colloid Science. 2nd edition, Oxford University Press, New York.

Israelachvili, J.N. (2011) Intermolecular and Surface Forces. 3rd ed., Elsevier, Academic Press, Amsterdam.

Israelachvili, J.N., Ninham, B.W. (1977) Intermolecular forces – the long and short of it. Journal of Colloid and Interface Science 58, 14–25. https://doi.org/10.1016/0021-9797(77)90367-8

IUPAC (2006) Compendium of Chemical Terminology, 2nd ed. (the "Gold Book") (1997). Online corrected version: (2006) "solvation". https://goldbook.iupac.org/terms/view/S05747

Jensen, W.B. (2016) Collected Papers, Volume 3, Acid-Base Chemistry and Related Topics. Oesper Collections, Cincinnati, Ohio, USA. https://www.google.com/url?sa=t&source=web&rct=j&opi=89978449&url=https://homepages.uc.edu/~jensenwb/books/Collected%2520Papers%2520Vol.%25203.pdf&ved=2ahUKEwjhmo7h8I2JAxWwZ_EDHfHoAfUQFnoECBoQAQ&usg=AOvVaw2j__bm5YQe1bfhsSsgo6_S

Jones, J.E. (1924a) On the determination of molecular fields – I. From the variation of the viscosity of a gas with temperature. Proceedings of the Royal Society of London. Series A, Containing Papers of a Mathematical and Physical Character 106 (738), 441–462. https://doi.org/10.1098/rspa.1924.0081

Jones, J.E. (1924b) On the determination of molecular fields – II. From the equation of state of a gas. Proceedings of the Royal Society of London. Series A, Containing Papers of a Mathematical and Physical Character 106 (738), 463–477. https://doi.org/10.1098/rspa.1924.0082

Keesom, W.H. (1915) The second viral coefficient for rigid spherical molecules, whose mutual attraction is equivalent to that of a quadruplet placed at their centre. Proceedings of the Royal Netherlands Academy of Arts and Science 18, 636–646. https://www.google.com/url?sa=t&source=web&rct=j&opi=89978449&url=https://dwc.knaw.nl/DL/publications/PU00012540.pdf&ved=2ahUKEwj0t6aLho6JAxU9SfEDHYGhJF8QFnoECBwQAQ&usg=AOvVaw0Yr9ezyg5Cd7Yaj7mXiz8p

Kékicheff, P. (2019) The long-range attraction between hydrophobic macroscopic surfaces. Advances in Colloid and Interface Science 270, 191–215. https://doi.org/10.1016/j.cis.2019.06.004

Khan, R., Latha, G.M. (2023) Multi-scale understanding of sand-geosynthetic interface shear response through Micro-CT and shear band analysis. Geotextiles and Geomembranes 51, 437–453. https://doi.org/10.1016/j.geotexmem.2023.01.006

Krumbein, W.C. (1941) Measurement and geological significance of shape and roundness of sedimentary particles. Journal of Sedimentary Research 11, 64–72. https://doi.org/10.1306/d42690f3-2b26-11d7-8648000102c1865d

Laskowski, J., Kitchener, J.A. (1969) The hydrophilic–hydrophobic transition on silica. Journal of Colloid and Interface Science 29, 670–679. https://doi.org/10.1016/0021-9797(69)90219-7

Laurence, C., Gal, J.-F. (2010) Lewis Basicity and Affinity Scales. Data and Measurement. Wiley.

Lee, H.M., Kim, Y.W., Go, E.M., Revadekar, C., Choi, K.H., Cho, Y., Kwak, S.K., Park, B.J. (2023) Direct measurements of the colloidal Debye force. Nature Communications 14, 3838. https://www.nature.com/articles/s41467-023-39561-8

Lekkerkerker, H.N.W., Tuinier, R. (2011) Colloids and the Depletion Interaction. Springer.

Lekkerkerker, H.N.W., Tuinier, R., Vis, M. (2024) Colloids and the Depletion Interaction. 2nd ed., Springer (open access).

Lenhard, J., Stephan, S., Hasse, H. (2024) On the history of the Lennard-Jones potential. Annalen der Physik 536, 2400115. https://doi.org/10.1002/andp.202400115

Lewis, G.N. (1916) The atom and the molecule. Journal of the American Chemical Society 38, 762–785. https://doi.org/10.1021/ja02261a002

Lewis, G.N. (1923) Valence and the Structure of Atoms and Molecules, Chemical Catalog Company, New York, 141–142; reprinted as Lewis, G.N. (1966) Valence and the Structure of Atoms and Molecules, Dover: New York, NY, p. 141.

Lewis, G.N. (1938) Acids and bases. Journal of the Franklin Institute 226, 293–313. https://doi.org/10.1016/S0016-0032(38)91691-6

Lifshitz, E.M. (1956) The theory of molecular attractive forces between solids. Soviet Physics 2, 73–83. https://www.google.com/url?sa=t&source=web&rct=j&opi=89978449&url=https://www.mit.edu/~kardar/research/seminars/Casimir/LifshitzTheory.pdf&ved=2ahUKEwiRw5-Zh46JAxV8evEDHTyYCR0QFnoECBQQAQ&usg=AOvVaw37XHX1BMTGBz9z59EJkqTP

Lin, T., Lu, Z., Chen, W. (2014) Interaction mechanisms and predictions on membrane fouling in an ultrafiltration system, using the XDLVO approach. Journal of Membrane Science 461, 49–58. https://doi.org/10.1016/j.memsci.2014.03.022

Liu, J., Shen, L., Lin, H., Huang, Z., Hong, H., Chen, C. (2022) Preparation of Ni@UiO-66 incorporated polyethersulfone (PES) membrane by magnetic field assisted strategy to improve permeability and photocatalytic self-cleaning ability. Journal of Colloid and Interface Science 618, 483–495. https://doi.org/10.1016/j.jcis.2022.03.106

London, F. (1930) Zur Theorie und Systematik der Molekularkräfte. Zeitschrift für Physik 63, 245–279. https://doi.org/10.1007/BF01421741

Mahanty, J., Ninham, B.W. (1976) Dispersion Forces. Academic, London.

McClellan, A.L. (1963) Tables of Experimental Dipole Moments. W.H. Freeman & Co., San Francisco and London.

Mie, G. (1903) Zur kinetischen Theorie der einatomigen Körper. Annalen der Physik 316, 657–697. https://doi.org/10.1002/andp.19033160802

Miyazaki, K., Schweizer, K. S., Thirumalai, D., Tuinier, R., Zaccarelli, E. (2022). The Asakura-Oosawa theory: Entropic forces in physics, biology, and soft matter. Journal of Chemical Physics 156, 080401. https://doi.org/10.1063/5.0085965

Ninham, B.W. (1999) On progress in forces since the DLVO theory. Advances in Colloid and Interface Science 83, 1–17. https://doi.org/10.1016/S0001-8686(99)00008-1

Ninham, B.W. (2019) B. V. Derjaguin and J. Theo. G. Overbeek. Their times, and ours. Substantia 3, 65–72. https://doi.org/10.13128/Substantia-637

Ohki, S., Ohshima, H. (1999) Interaction and aggregation of lipid vesicles (DLVO theory versus modified DLVO theory). Colloids and Surfaces B: Biointerfaces 14, 27–45. https://doi.org/10.1016/S0927-7765(99)00022-3

Oosawa, F. (2021) The history of the birth of the Asakura–Oosawa theory. Journal of Chemical Physics 155, 084104. https://doi.org/10.1063/5.0049350

Parsegian, V.A. (2006) Van der Waals Forces. A Handbook for Biologists, Chemists, Engineers, and Physicists. Cambridge University Press, Cambridge.

Prieve, D.C. (1999) Measurement of colloidal forces with TIRM. Advances in Colloid and Interface Science 82, 93–125. https://doi.org/10.1016/S0001-8686(99)00012-3

Stern, O. (1924) Zur Theorie der elektrolytischen Doppelschicht. Zeitschrift für Elektrochemie 30, 508–516. https://doi.org/10.1002/bbpc.192400182

Stumm, W., Morgan, J.J. (1996) Aquatic Chemistry. Chemical Equilibria and Rates in Natural Waters. 3rd ed., Wiley & Sons, New York.

Tadros, T.F. (2010) Rheology of Dispersions. Principles and Applications. Wiley-VCH, Weinheim.

Thormann, E. (2017) Surface forces between rough and topographically structured interfaces. Current Opinion in Colloid & Interface Science 27, 18–24. https://doi.org/10.1016/j.cocis.2016.09.011

van Oss, C.J. (1993) Acid-base interfacial interactions in aqueous media. Colloids and Surfaces A: Physicochemical and Engineering Aspects 78, 1–49. https://doi.org/10.1016/0927-7757(93)80308-2

van Oss, C.J. (1994) Interfacial Forces in Aqueous Media, Marcel Dekker, New York, 2nd ed. 2020, CRC Press.

van Oss, C.J. (2003) Long-range and short-range mechanisms of hydrophobic attraction and hydrophilic repulsion in specific and aspecific interactions. Journal of Molecular Recognition 16, 177–190. https://doi.org/10.1002/jmr.618

van Oss, C.J. (2008) Chapter Three – The Extended DLVO Theory, Interface Science and Technology, Volume 16, 31–48. https://doi.org/10.1016/S1573-4285(08)00203-2

Verwey, E.J.W., Overbeek, J.T.G. (1948) Theory of Stability of Lyophobic Colloids. The Interaction of Sol Particles Having an Electric Double Layer. Elsevier Publishing Company, New York – Amsterdam – London – Brussels.

Wadell, H (1932) Volume, shape, and roundness of rock particles. Journal of Geology 40, 443–451. http://dx.doi.org/10.1086/623964

Walz, J.Y. (1998) The effect of surface heterogeneities on colloidal forces. Advances in Colloid and Interface Science 74, 119–168. https://doi.org/10.1016/S0001-8686(97)00042-0

Wang, S., Lia, L., Yu, S., Dong, B., Gao, N., Wang, X. (2021) A review of advances in EDCs and PhACs removal by nanofiltration: Mechanisms, impact factors and the influence of organic matter. Chemical Engineering Journal 406, 126722. https://doi.org/10.1016/j.cej.2020.126722

Wentworth, C.K. (1922) The shapes of beach pebbles. U.S. Geological Survey professional paper 131C, 75–83. https://doi.org/10.3133/pp131C

Zhao, L., Schwarz, W.H.E., Frenking, G. (2019) The Lewis electron-pair bonding model: the physical background, one century later. Nature Reviews Chemistry 3, 35–47. https://doi.org/10.1038/s41570-018-0052-4

Zingg, T. (1935) Beitrag zur Schotteranalyse. Schweizerische mineralogische und petrographische Mitteilungen 15, 39–140. https://www.research-collection.ethz.ch/handle/20.500.11850/135183

Chapter 2
Destabilisation of colloid systems – its initiation

Two opposite terms – repulsion and attraction of the particles – strictly separate two branches of industrial applications dealing with the colloidal particles. One branch based on optimally homogeneous distribution of colloidal particles (cosmetology, painting, pharmaceuticals, etc.) is interested in repulsive forces, and the other one ((waste)water treatment, etc.) studies how to make the attractive forces dominating with the aim to aggregate and consequently separate the particles from the carrier liquids.

The DLVO (for non-aqueous systems) and XDLVO (for aqueous systems) theories presented in the preceding chapter indicate the stages of full stability (separation of the individual particles) and the conditions under which the systems are susceptible to initial destabilisation. Under these conditions an application of the coagulants and additive agents participating in floc formation makes this process more efficient.

The ability of a dispersion to resist cluster formation is denoted as colloid stability which is of kinetic (a force barrier preventing collisions between the colloidal particles) or thermodynamic (preventing an increase in free energy) nature. When this stability is disturbed, in other words, a process of coagulation is initiated, and the individual particles are packed together (see Fig. 2.1). Already at this stage it is apparent how geometrical shapes of the particles participate in compactness of the forming aggregates. The spherical particles are much more predisposed to the consequent breakage under the applied shearing and new restructuring in a comparison with the irregular particles which may be mutually locked.

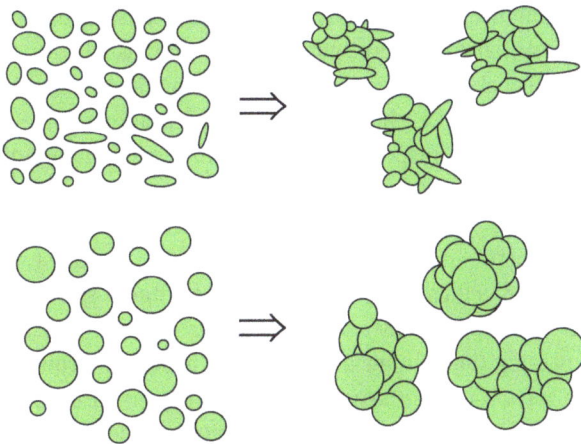

Fig. 2.1: A difference between formation of spherical and irregular particle clusters in the process of destabilisation.

https://doi.org/10.1515/9783111246765-002

Terminology of two tightly related notations – coagulation and flocculation – is not, in general, strictly delimited and practical meanings of these two words are in many respects interlaced with a rather blurred boundary. Coagulation is related with an initial phase of floc formation. However, the term coagulation does not represent only an initial stage of particles attraction and forming of primary clusters but covers the whole process including an appropriate choice of coagulants enhancing clusters creation, adequate dosage of coagulants, and especially an optimisation of pH values under which coagulation and consequently flocculation processes are conducted.

In the following, some aspects, how to start and facilitate a process of coagulation (charge neutralisation), are presented. First, the notion of the ζ-potential and its evaluation using electrokinetic phenomena is discussed. The consequent section is devoted to the Schulze-Hardy rule indicating which coagulants should be efficient. The next section presents a basic categorisation of the coagulants, and finally the last section is devoted to the evaluation of coagulant dosage.

2.1 Destabilisation of colloid systems – its evaluation

2.1.1 Zeta (ζ) potential

Helmholtz (1853) was the first who realised that a charged electrode in the electrolyte solution repels the co-ions of the same charge from its surface and simultaneously attracts the counterions of the opposite charge. In this way two layers of opposite polarity create an electrode-electrolyte interface.

A dependence of capacitance on the applied potential and the ionic concentration directed Gouy (1910) and Chapman (1913) to an introduction of the so-called diffuse layer not so compact as in the Helmholtz case with fully or nearly fully adsorbed ions. This enabled to describe an exponential decrease of potential with the distance from the interface. Interactions of ions (considered as point charges which polarisability is neglected) are supposed to be of purely electrostatic nature (no specific adsorption).

A problem with highly charged layers inspired Stern (1924) to a combination of the two preceding approaches: a fixed (nowadays called) Stern layer of counterions adsorbed to the interface followed by a diffuse layer. The Stern layer was modified by Grahame (1947) who distinguished between two types of counterions in the closest vicinity of the interface. The first group is formed by the counterions without solvation shells with a direct contact with ions ('specifically adsorbed counterions') with the so-called inner Helmholtz plane passing the counterions centres. The second group is formed by the counterions with solvation shells separating them from the direct contact with the surface particle ions but still with a very strong bound to the interface. The plane passing very closely to the centres of these counterions is denoted as the outer Helmholtz one. This second Helmholtz layer then converts to the Gouy-Chapman diffuse layer gradually balancing the ions and counterions presence to a level corresponding to the bulk region (bulk solution) (see Fig. 2.2). Hence, the notion electrical double layer (EDL) covers the charged surface together with a diffuse layer. More details are introduced, for instance, in the monographs by Hiemenz and Rajagopalan (1997) and Israelachvili (2011).

The zeta (ζ) potential (depicted in Fig. 2.2) represents a fundamental characteristic in behaviour of colloidal particles (Hunter 1981). Its value is given by the potential difference between a shear (slipping) plane (separating the immobile and mobile layers) and a passage from a diffuse layer to a bulk fluid and is always lower (in an absolute value) than the Stern potential ψ_δ. An advantage of the zeta potential over the total (surface) ψ_0 and Stern ψ_δ potentials underlying its importance is a possibility of its evaluation by the experimental technique. At most cases this is the only way how to characterise an EDL. The absolute values of the zeta potential indicate an (im) balance between the attractive van der Waals forces and electrostatic repulsive forces (see Fig. 2.3). The higher values ensure stability of a colloid system, with lowering of ζ value coagulation of particles becomes more probable. The zeta potential can be considered as more relevant for the description of structure stability as double layer interactions between the particles primarily depend on behaviour of the diffuse layers forming the hulls round particles. The ζ-potential is not an absolutely independent quantity but strongly subjects to pH and ionic strength. More information on the ζ-potential is given in Section 2.1.6.

2.1.2 Water density distribution function

Water density distribution function ranges among the factors characterising (to some minimal input) the EDL region. The behaviour of this function is intensively studied at the metal-water interfaces (see, for instance, Sakong and Gross 2018). Tesch et al. (2021) applied the computational method based on the interaction potential of Lennard-Jones type (see eq. (1.38)) completed with a separate term for Coulomb interactions. In this way they obtained an oscillatory course of water density distribution function.

Lyklema (2010) predicted the same behaviour (see Fig. 2.4). It can be suspected that a passage from the stagnant layer of EDL attached to the solid surface to the movable part (a diffuse layer) is at the location of a local water density minimum. This could indicate a location of the shear plane, and hence, a location of the zeta potential – the closest experimentally measurable magnitude to the surface along the potential curve.

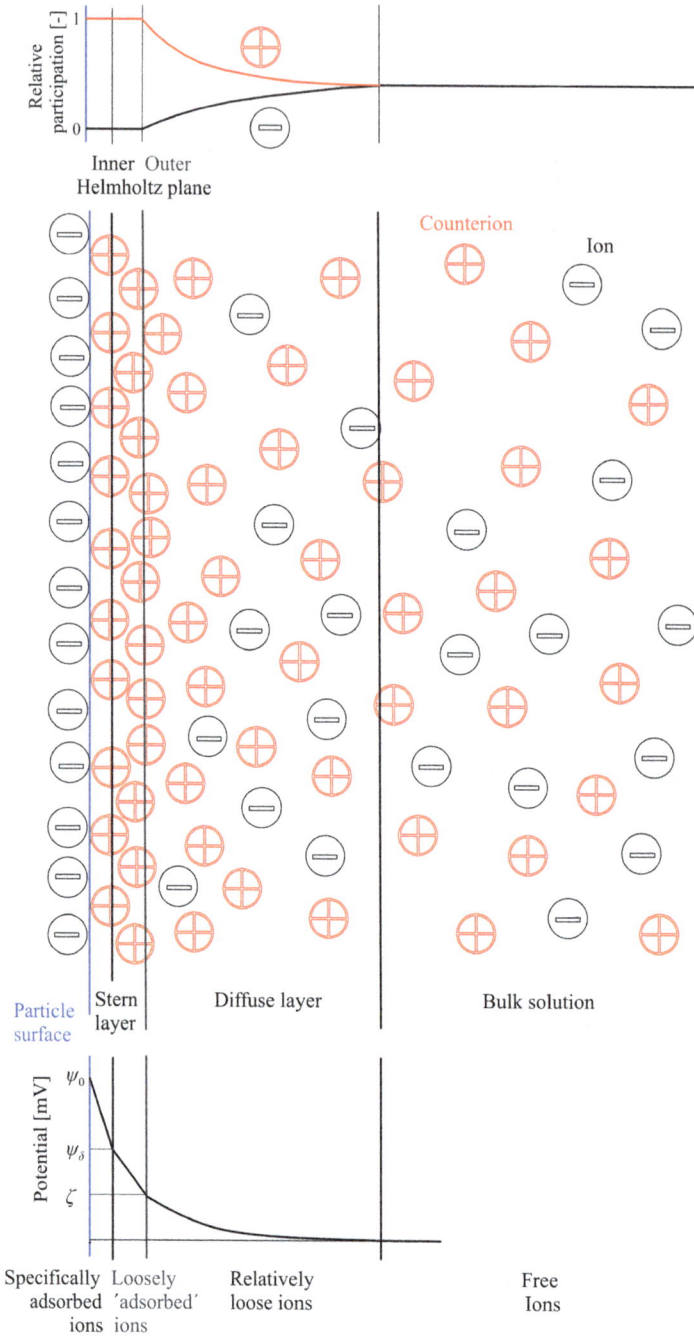

Fig. 2.2: Arrangement of the electrical double layer with the courses of zeta potential and ion participation (anions vs. cations).

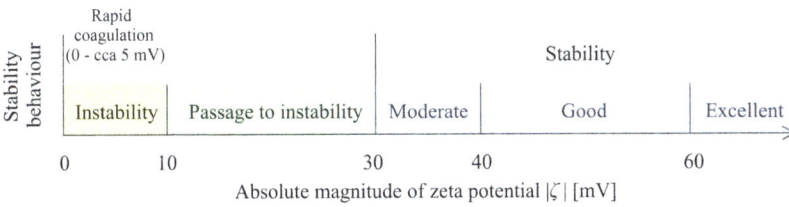

Fig. 2.3: (In)stability behaviour of a colloid in dependence on the values of the ζ-potential.

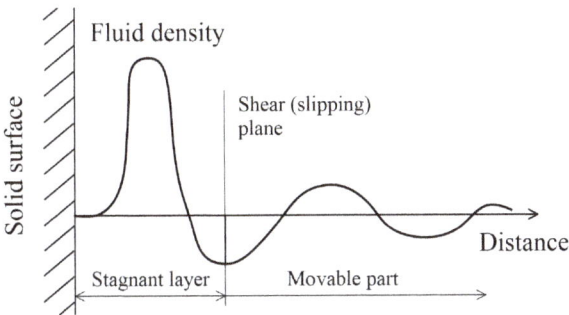

Fig. 2.4: The oscillatory character of water density distribution function.

2.1.3 Temperature effects on a rate of coagulation

Final tuning of specific coagulant(s), its (their) dosage, and an appropriate choice of pH value in the individual water treatment plants is only temporary (if we exclude such factors as heavy rains and others) – at most seasonal – as a change of temperature reflects in a rearrangement of these entry factors. The reason is simple: temperature ranges among the most crucial parameters in the coagulation process and most physical quantities – such as viscosity, density, diffusivity, electric conductivity, permittivity, intensity of Brownian motion – are more or less temperature-dependent. This has an immediate impact also to a chemical aspect of the coagulation process and especially to its rate and floc morphology and compactness.

Camp et al. (1940) attributed seasonal changes in water treatment dominantly to the effect of viscosity on sedimentation. If we compare shear viscosity of water at 5 °C ($\approx$1.52 mPa s) against its value at 25 °C ($\approx$0.89 mPa s), see eq. (1.6) and Fig. 2.5, then water resistance against flow (viscosity) at 5 °C is by 71% higher. However, this cannot be taken as the only reason.

Changes of water density ρ_w (5 °C – 1.0000 g cm^{-3}, 25 °C – 0.9970 g cm^{-3}) are negligible in contrast to water relative permittivity ε_r (5 °C – 85.8, 25 °C – 78.3) expressing a degree of suspension polarity (see Fig. 2.6). An increase at 5 °C attains approximately

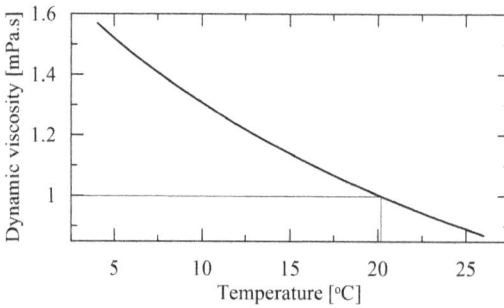

Fig. 2.5: Dependence of dynamic viscosity of water on temperature.

9.5% (see Fig. 2.7) resulting from a relatively accurate approximation by Malmberg and Maryott (1956):

$$\varepsilon_r = 87.740 - 0.40008 \times Temp + 9.398 \times 10^{-4} \times Temp^2 - 1.410 \times 10^{-6} \times Temp^3, \qquad (2.1)$$

where *Temp* is the temperature at °C. A more detailed analysis of permittivity data is presented by Ellison (2007).

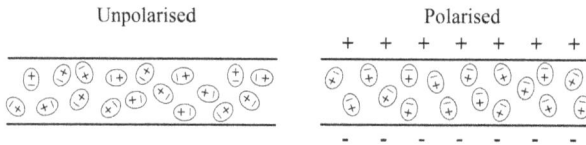

Fig. 2.6: Permittivity – a measure of the electric polarisability of dielectric materials.

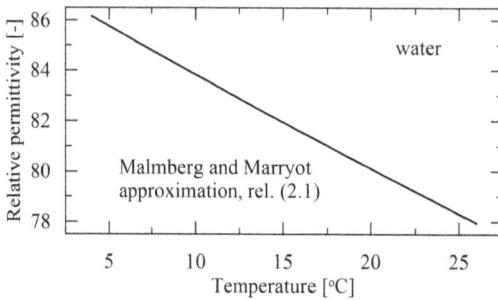

Fig. 2.7: Dependence of relative permittivity of water on temperature.

Generally, it is possible to say that solubility – the ability of a solute to form a solution with the solvent – increases with temperature (in the cases of solid and liquid solutes in the liquid solvents). However, a course of the amount of a solute to form a saturated solution is individual for various couples (solute-solvent). The saturated amount of a solute is usually given in g/100 g of solvent. Nevertheless, presentation of solubility in moles instead of mass (mol/kg) provides better comparison in usage (consumption) of various coagulants and their efficiencies.

Intensity of Brownian motion substantially contributes to successfulness of the process of coagulation. Its dependence on temperature is really non-negligible as documented in Table 1.1. The mean displacement of particles subjected to their dimensions (diameters) decreases in tens of % with a passage of water temperature from 25 to 5 °C. This is one of the main causes significantly contributing to reduction of a coagulation process at lower temperatures as low temperature implies low collision frequency (Yu et al. 2012).

The above short list of inputs deteriorating a rate of coagulation with a decreasing temperature participates in all-year-round problems with quality of treated water. The detrimental influence of reduced temperature was presented by Kang (1995), Xiao et al. (2008), and Xiao et al. (2009). The degree of reduced coagulation with decreasing temperature is not uniform for the individual coagulants and substantially differs. As an example, it is possible to introduce aluminium as unsuitable coagulant at lower temperatures (see Morris and Knocke 1984).

Lower temperatures always result in the creation of reduced floc size and the formation of weaker flocs (Morris and Knocke 1984; Haarhoff and Cleasby 1988; Hanson and Cleasby 1990). Changes in morphology also project to a characterisation of flocs by means of fractal dimension (see Section 4.4.5).

Occasionally, the situation with temperature drop can be handled by increasing pH value (Camp et al. 1940; Mohtadi and Rao 1973; Hanson and Cleasby 1990). Turbidity removal efficiency in dependence on temperature is another apparent problem (Dayarathne et al. 2022).

The problems with coagulation are not restricted only to temperatures approaching lower values (2 and 5 °C) but also arises in the opposite direction exceeding 30 °C as documented for 40 °C by Dayarathne et al. (2022).

2.1.4 Electrokinetic phenomena

The values of potentials in the EDL (related to the potential of a bulk region) are theoretically well-defined (surface potential ψ_0 and Stern potential ψ_δ); however the only measurable potential is the ζ-potential (also called the electrokinetic potential) attaining its value at the shear (slipping) plane and indicating a measure of (de)stability of the colloid system. This value can be measured by means of the so-called electrokinetic phenomena. The term electrokinetic is related to a movement of charged particles in liquid medium exposed to an electric field. There are four traditional electrokinetic phenomena based on a combination of applied mechanical and electric forces, each characterised by a corresponding relative tangential motion between an electrolyte solution (or its component) and a charged solid surface. An electric force applied in two phenomena – electroosmosis and electrophoresis – induces a mechanical motion, and conversely an applied mechanical force in the other two phenomena – streaming potential and sedimentation potential – produces an electric current (potential). The cause in the first case is obvious: an electrically charged particle dissolved in aqueous solution exposed to an electric field migrates towards the electrode bearing the opposite charge. The cause in the other case is reciprocal: applied pressure gradient influences the characteristics in the hydrodynamic region (comprising a diffuse layer) and thus participates in the characteristics of the EDL. Specifically, the individual electrokinetic phenomena are generated as follows:

- Electroosmosis – an applied electric field results in a motion of electrolyte past a charged surface;
- Electrophoresis – an applied electric field results in a motion of particles relative to a stationary carrier liquid;
- Streaming potential – an application of pressure gradient creates the electric field (the opposite of electroosmosis);
- Sedimentation potential – sedimentation of charged particles in a carrier liquid evokes the electric field (the opposite of electrophoresis).

Hence, there are two couples of electrokinetic phenomena where the cause and effect are mutually interchanged.

Intensity of electrokinetic phenomena is closely related with rheological character (loosely said with a course of viscosity) of a carrier liquid. Dominantly an interest is devoted to Newtonian liquids such as water which viscosity depends only on pressure and temperature and not on shearing. Shear thinning liquids (viscosity decreases with an increasing shear rate) enhance electrokinetic phenomena in contrast to shear thickening liquids (viscosity increases with an increasing shear rate) for which the electrokinetic effects are reduced (Zhao and Yang 2013).

The individual electrokinetic phenomena are introduced in the following sections. Detailed analyses are presented in Hiemenz and Rajagopalan (1997, Chapter 12), Masliyah and Bhattacharjee (2006, Chapter 7), and Delgado et al. (2007).

2.1.4.1 Electroosmotic flow

An electric field applied to a solution contained in the capillary or between parallel plates evokes a motion of counterions dominantly in the diffuse layer where their concentration prevails. The counterions are accompanied by the surrounding hydrated shells (in the case, that a carrier liquid is water), thus jointly migrating towards an opposite electrode. The situation corresponding to the initial state is depicted in Fig. 2.8.

The velocity profile exhibits a shape of the two-humped camel where both humps are more progressive with increasing electric field and temperature. Increasing temperature of a carrier liquid contributes to non-negligible decrease in a resistance against flow as documented in Tab. 2.1. A decrease in kinematic viscosity (a ratio of

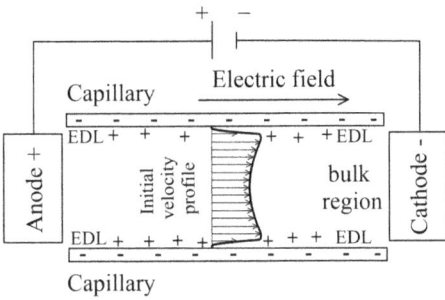

Fig. 2.8: The initial velocity profile of a suspension flow under an application of electric field.

dynamic viscosity and fluid density) reflects in an increase of the Reynolds number (non-dimensional number relating inertial and viscous forces) which accentuates an occurrence of the humps. This is in correspondence with the findings in Yang et al. (2001). During a relatively short time instant the whole volume is uniformly dragged, the velocity profile exhibits a flat head tending to zero velocity at the close proximity to the charged walls (the inner Helmholtz layer with the adsorbed immobile counterions) (see Fig. 2.9). The width of this immobile layer is under normal condition lower than 1 nm in contrast to the diffuse layer attaining several nanometres or even micrometres.

Tab. 2.1: A comparison of the magnitudes of kinematic viscosity of water at various temperatures.

Temperature (°C)	5	10	15	20	25
Kinematic viscosity (mm^2 s^{-1})	1.52	1.31	1.14	1.00	0.89

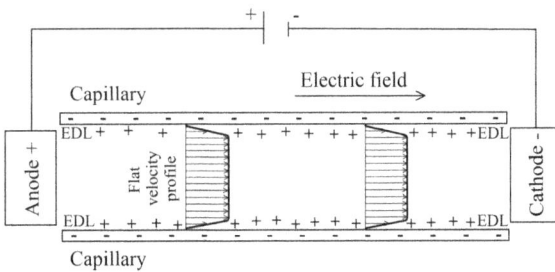

Fig. 2.9: The flat velocity profile of the electroosmotic flow (with exaggerated wings) following the two-humped one.

The width of the diffuse layer (as a dominating part of the EDL) is approximated by the so-called Debye length λ_D (related to the Debye-Hückel parameter κ by $\kappa = \lambda_D^{-1}$)

$$\lambda_D = \left(\frac{\varepsilon_r \varepsilon_0 k_B T}{2 z^2 e^2 c_0} \right)^{1/2} \tag{2.2}$$

where ε_r is the relative permittivity and ε_0 is the vacuum permittivity (= 8.8541878188 (14) $\times 10^{-12}$ F/m), k_B is the Boltzmann constant (= 1.380649 $\times 10^{-23}$ J/K), T is the absolute temperature, z is the ionic valence, e is the elementary charge (=1.602176634 $\times 10^{-19}$ C), and c_0 is the electrolyte concentration. A course of concentration of counterions can be characterised by an exponential decrease across the diffuse layer, thus enabling an appearance of the two humps at the beginning of the whole process.

The above-described flow based on the existence of EDL and an application of an electric field is called an electroosmotic flow and was first presented already in 1807 (published 2 years later, Reuss 1809). As electroosmotic flow is a surface-driven phenomenon, its velocity is practically independent on a capillary diameter or a channel width. Among its disadvantages it is possible to name potential surface irregularities, surface contamination, a requirement of relatively high electric field (exceeding 100 V/cm), and resulting moderate flow velocity (usually not attaining 1 mm/s).

The evaluation of the magnitude of the flat velocity profile (see Fig. 2.9) can be derived by the so-called Helmholtz-Smoluchowski approach (Smoluchowski 1903). As usual, in derivation it is necessary to start with the Navier-Stokes balance equation

$$\rho \frac{D\mathbf{u}}{Dt} = \nabla \cdot \mathbf{T} + \rho g + \rho_E \mathbf{E} \tag{2.3}$$

where ρ is the fluid density, $\mathbf{u}$ is the velocity vector, D/Dt denotes the material derivative (the time rate of change of fluid velocity), $\mathbf{T}$ is the stress tensor, g is the gravitational acceleration (=9.80665 m/s^2), ρ_E is the electric charge density, and $\mathbf{E}$ is the electric field vector. If we suppose inertia-free capillary flow with no pressure gradient and if we neglect gravitational forces (colloid system), then the above equation can be substantially simplified, which results in the balance equation between viscous and electrical forces

$$\mu \nabla^2 \mathbf{u} = -\rho_E \mathbf{E}. \tag{2.4}$$

Now we can apply the order-of-magnitude approach (neglecting less importance coordinates and components); in this case we can neglect the derivatives with respect to the longitudinal coordinate x directed towards the electrode and suppose that a velocity depends only on the transverse coordinate y. This implies another simplification and together with the assumption that the width of the diffuse layer λ_D is small in comparison with the channel dimension width (enabling a suppression of the curvature terms) we transform the originally partial differential equations to the one-dimensional ordinary one

$$\mu \frac{\partial^2 u}{\partial y^2} = -\rho_E E_x, \tag{2.5}$$

where E_x denotes the component of the applied electric field vector in the longitudinal direction x.

In the next step, the electric charge density ρ_E can be expressed through the Poisson equation

$$\nabla^2 \psi = -\frac{\rho_E}{\varepsilon} \qquad (2.6)$$

relating potential, electric charge density, and permittivity. Hence, we get

$$\mu \frac{\partial^2 u}{\partial y^2} = \varepsilon \frac{\partial^2 \psi}{\partial y^2} E_x. \qquad (2.7)$$

Supposing the boundary conditions $\partial u/\partial y = \partial\psi/\partial y = 0$ at the boundary of the diffuse layer (constant profiles of u and ψ in the bulk region) and integrating the preceding equation we obtain

$$\mu \frac{\partial u}{\partial y} = \varepsilon \frac{\partial \psi}{\partial y} E_x. \qquad (2.8)$$

Integrating once more and substituting the zeta potential ζ for ψ at $u = 0$, we finally derive the Helmholtz-Smoluchowski relation (past a planar charged surface) for the longitudinal developed velocity component (flat profile) of the electroosmotic flow in the form

$$u = -\frac{\varepsilon_r \varepsilon_0 \zeta E_x}{\mu}, \qquad (2.9)$$

which is independent on the channel width. The negative sign corresponds to the introduction of electric potential ψ. For negatively charged ions at the surface the counterions have a positive charge, a sign of the zeta potential ζ is negative, and hence, with respect to eq. (2.9) the velocity is directed to the cathode.

The term

$$\mu_E = \frac{\varepsilon_r \varepsilon_0 \zeta}{\mu} \qquad (2.10)$$

relating the electroosmotic velocity and the applied electric field (eq. (2.9)) is denoted as the electroosmotic mobility.

Smoluchowski (1903) derived eq. (2.9) for the case of relatively large particles ($\kappa a \gg 1$, where a is the particle radius). Later on, Debye and Hückel (1924) incorporated electrophoretic retardation and frictional drag of the particle (both factors reduce mobility) and derived the relation between the electroosmotic velocity and the applied electric field in the form (nowadays called the Hückel-Onsager relation)

$$u = -\frac{2}{3}\frac{\varepsilon_r \varepsilon_0 \zeta E_x}{\mu}, \qquad (2.11)$$

in this case for small particles ($\kappa a \ll 1$). This seeming discordance with eq. (2.9) was harmonised by Henry (1931) introducing the relation

$$u = -\frac{2}{3}\frac{\varepsilon_r\varepsilon_0\zeta E_x}{\mu}f(\kappa a),$$
(2.12)

where the correction factor $f(\kappa a)$ takes into consideration the particle shape (for spherical particles and low potentials monotonous f attains the values from 1 to 3/2 for κa varying between the limiting values from 0 to ∞ (see also Hunter (1981, Sect. 3.3.1), Elimelech et al. (1995, Sect. 2.3.4), and Hiemenz and Rajagopalan (1997, Sect. 12.5a). The expressions for $f(\kappa a)$ (supposing a non-conducting sphere and assuming that the external field deformed by the presence of the colloidal particle and the field of the double layer are additive) are of the form (Henry 1931)

$$f(\kappa a) = 1 + \frac{1}{16}(\kappa a)^2 - \frac{5}{48}(\kappa a)^3 - \frac{1}{96}(\kappa a)^4 + \frac{1}{96}(\kappa a)^5 -$$
$$- \frac{11}{16}(\kappa a)^6 \times \exp(\kappa a) \times \int_{\infty}^{\kappa a}\frac{e^{-t}}{t}dt \quad for \; \kappa a < 5$$
(2.13a)

$$f(\kappa a) = \frac{3}{2} - \frac{9}{2}(\kappa a)^{-1} + \frac{75}{2}(\kappa a)^{-2} - 330(\kappa a)^{-3} + ... \; for \; \kappa a > 25.$$
(2.13b)

Each relation attains the corresponding limiting value, i.e. eq. (2.13a) approaches 1 for decreasing values of κa, eq. (2.13b) tends to 3/2 for increasing values of κa. As Henry (1931) introduces, there is a hiatus in the series of computations between 5 and 25. Ohshima (1994) spanned this shortcoming by proposing a function (after rearrangement)

$$f(\kappa a) = \frac{2}{3} + \frac{1}{3}\left[\frac{\kappa a(1+2e^{-\kappa a})}{2.5+\kappa a(1+2e^{-\kappa a})}\right]^3$$
(2.14)

smoothly passing from the limiting value 2/3 ($\kappa a \to 0$, the Hückel-Onsager equation, the term in the brackets tends to 0) to the other limiting value 1 ($\kappa a \to \infty$, the Helmholtz-Smoluchowski relation, the term in the brackets tends to 1). The deviation of this function from the original Henry's curve (both branches) does not exceed 1%.

The original Henry correction can also be approximated by the simpler expression

$$f(\kappa a) = \frac{5}{6} + \frac{1}{6}\tanh\left[\frac{4}{3}(\log(\kappa a)-1)\right]$$
(2.15)

providing practically the identical course of the Henry correction as the Ohshima's one (see Fig. 2.10).

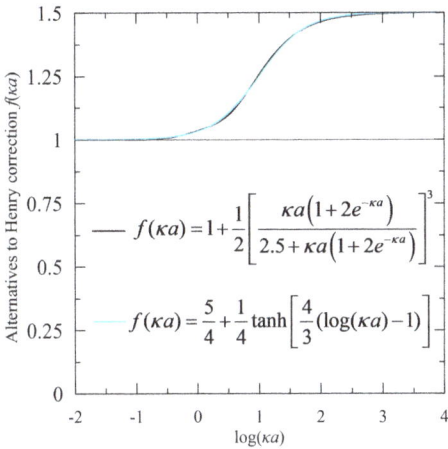

The graph shows:

$$f(\kappa a) = 1 + \frac{1}{2}\left[\frac{\kappa a\left(1 + 2e^{-\kappa a}\right)}{2.5 + \kappa a\left(1 + 2e^{-\kappa a}\right)}\right]^{3}$$

$$f(\kappa a) = \frac{5}{4} + \frac{1}{4}\tanh\left[\frac{4}{3}(\log(\kappa a) - 1)\right]$$

Fig. 2.10: A comparison of the courses of the correction function $f(\kappa a)$ proposed by Ohshima (1994) and the functional form given by rel. (2.15).

For higher values of potential, the function f does not exhibit monotonous course, first decreases (for lower values of κa) from the limiting value 1 and then increases to the limiting value 3/2 (see Wiersema et al. 1966).

It is necessary to emphasise the striking difference if the fluid flow is exerted by an application of electric field or pressure gradient (see Fig. 2.11). The velocity profile generated by pressure gradient is represented by the classical parabolic profile while an application of electric field results in the plug flow region.

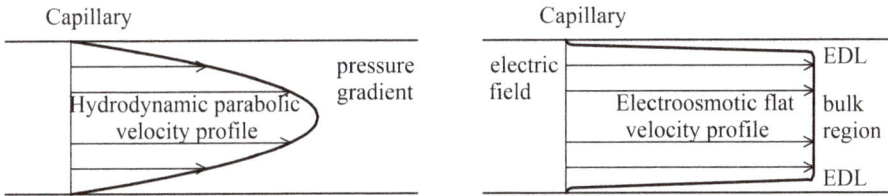

Fig. 2.11: Apparent difference in courses of fluid velocity profiles exerted by pressure gradient (left) and electric field (right).

2.1.4.2 Electrophoretic flow

In contrast to electroosmosis where – by applying an electric field – the liquid moves along the solid body, electrophoresis exhibits a motion of the dispersed particles with respect to the practically stagnant carrier liquid. The electrostatic force caused by an applied electric field is opposed by a frictional force including the forces generated by the EDL surrounding the particles. The positively charged particles move to the negative electrode and vice versa.

The electrophoretic velocity v_{ephor} is related with the applied electric field E by means of the electrophoretic mobility μ_{Ephor}

$$v_{\text{ephor}} = \mu_{\text{Ephor}} \times E \tag{2.16}$$

Using Smoluchowski's approach it is possible to derive (analogously to the case of electroosmosis) the relation for the electrophoretic mobility (however with the opposite sign), see rel. (2.12)

$$\mu_{\text{Ephor}} = -\frac{2}{3}\frac{\varepsilon_r \varepsilon_0 \zeta}{\mu} f(\kappa a), \tag{2.17}$$

where $f(\kappa a)$ is given by rel. (2.15)

$$f(\kappa a) = \frac{5}{6} + \frac{1}{6}\tan h\left[\frac{4}{3}(\log(\kappa a) - 1)\right].$$

Applying Ohm's law, the electric current I is expressed by

$$I = AK^{\text{bulk}}E, \tag{2.18}$$

where A is the capillary (microchannel) cross-sectional area and K^{bulk} is the bulk conductivity. Inserting E from the preceding relation into rel. (2.16) we obtain from the relation for volumetric flow rate $Q = Av_{\text{ephor}}$ the following expression relating the applied electric field with the zeta potential

$$\frac{Q}{I} = \frac{\varepsilon_r \varepsilon_0}{\eta K^{\text{bulk}}}\zeta. \tag{2.19}$$

2.1.4.3 Streaming potential

When a pressure gradient is applied to electrolyte suspension in a microchannel or a capillary with charged inner walls, the counterions in the mobile region of EDL are pushed downstream which establishes an electric current. This so-called streaming current I_{str} creates an electrokinetic potential, which, in turn, opposes the mechanical transfer of charge in the suspension flow direction and generates an electric current in the opposite direction called a conduction (leak) current I_c (see Fig. 2.12). At an equilibrium state both currents (streaming and conduction) are balanced, and the net current nullifies. At this stage the term streaming potential denotes the measured potential difference (see Fig. 2.13). It is necessary to remind that this electrokinetic phenomenon is generated only by an application of pressure gradient in the absence of an applied electric field.

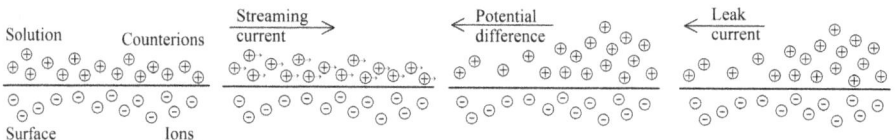

Fig. 2.12: Creation of the conduction (leak) current.

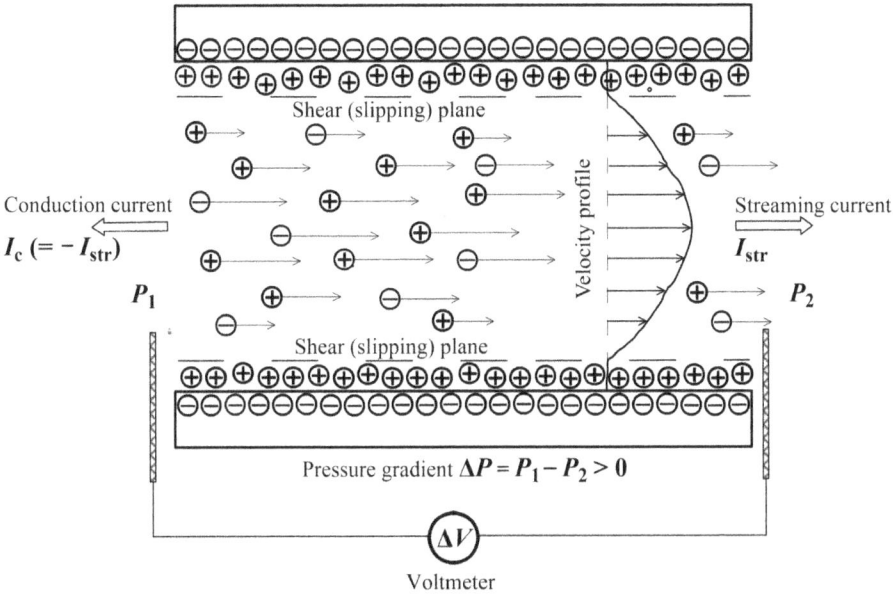

Fig. 2.13: Measurement of the streaming potential.

From the Poiseuille (describing laminar flow in a cylinder) and Poisson (describing solution charge density) equations it is possible to derive under the Smoluchowski's assumptions the relations for the streaming and conduction currents in the cylindrical capillary of a radius a and length L:

$$I_{str} = -\frac{\varepsilon_r \varepsilon_0 \pi a^2}{\eta} \frac{\Delta p}{L} \zeta, \tag{2.20}$$

$$I_c = \pi a^2 K^{bulk} \frac{U_{str}}{L}, \tag{2.21}$$

where K^{bulk} is the bulk conductivity and U_{str} is the streaming potential.

As $I_{str} + I_c = 0$, we obtain the relation between the streaming potential and the zeta potential in the form

$$U_{str} = \frac{\varepsilon_r \varepsilon_0}{\eta} \frac{\Delta p}{K^{bulk}} \zeta. \tag{2.22}$$

This derivation is only illustrative. It is necessary to have in mind under which conditions this relation was derived. The more general relations valid under weaker conditions can be found in recent journal publications.

2.1.4.4 Sedimentation potential

During sedimentation of the charged particles the equilibrium symmetry of their EDLs is distorted. Due to a motion, a regular spherical EDL boundary between a diffuse region and bulk volume is 'elliptically' prolonged which causes a cumulation of counterions in a 'free' diffuse volume (see Fig. 2.14). The lower part (taken from the sedimentation orientation) of the diffuse layer of the particle is continuously supplied by the counterions from the bulk volume and consequently returned back to the bulk volume at the prolonged upper part. This reflects in a displacement between the surface charge and the electric charge of the diffuse layer. In this way the moving particles create dipole moments (Booth 1954). The total sum over all particles generates an electric field E_{sed} for which the term sedimentation potential is used. This can be measured using two electrodes placed in the upper and bottom parts of the vessel.

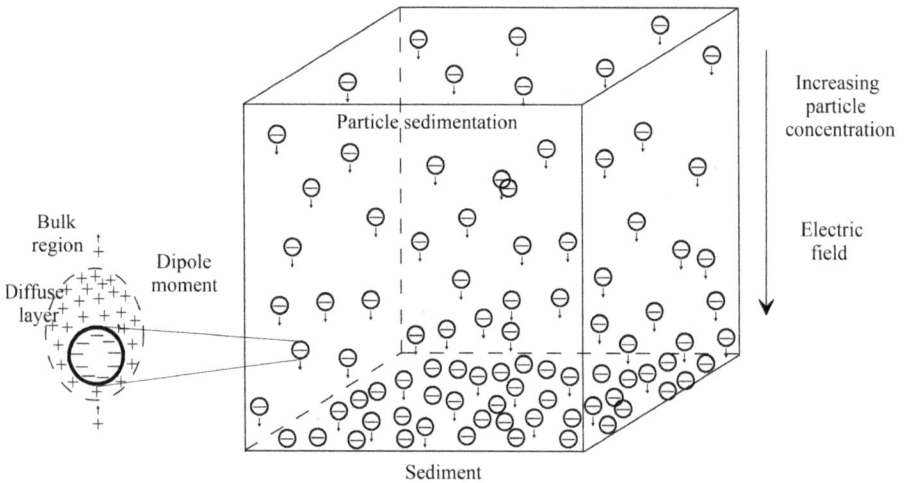

Fig. 2.14: The principle of the sedimentation potential.

The linear dependence of electric field on the zeta potential is given by Smoluchowski (1921)

$$E_{sed} = -\frac{\varepsilon_r \varepsilon_0 (\rho - \rho_0) \phi_p g}{K^{bulk} \eta} \zeta \tag{2.23}$$

where ε_r is the relative permittivity of medium, ε_0 is the permittivity of free space, ρ is the density of the particle, ρ_0 is the density of the medium, and ϕ_p is the particle volume fraction ($=(4\pi a^3 N_p)/(3\ V)$, where N_p is the particle number), g is the acceleration due to gravity, K^{bulk} is the bulk conductivity, and η is the viscosity.

Nevertheless, it is necessary to have in mind that this relation was derived under the simplifying assumptions (summarised in Marlow and Rowell 1985) that the monodispersed spherical sufficiently large ($\kappa a \gg 1$) particles are nonconducting, interparticle interactions are negligible, and the Reynolds number is lower than 1 (laminar flow). The adjacent EDLs are not overlapping.

The above-introduced Smoluchowski's relation was generalised by Ohshima et al. (1984). In accordance with Booth (1954) they did not put any restrictions on the particle sizes and EDL thicknesses but in contrast to Booth (1954) they do not suppose the low values of ζ. Moreover, they derived the explicit analytical relations for E_{sed} for two cases (low ζ values and large κa), where the right-hand side in rel. (2.23) is multiplied by the correction functions. Simultaneously, for both cases, they obtained the Onsager's reciprocal relation for sedimentation potential and electrophoresis represented by the sedimentation potential E_{sed} and electrophoretic mobility μ_E

$$E_{sed} = -\frac{(\rho - \rho_0)\phi_p g}{K^\infty}\mu_E, \tag{2.24}$$

where

$$K^\infty = \sum_{i=1}^{N} z_i^2 e^2 n_i^\infty / \lambda_i, \tag{2.25}$$

λ_i is the drag coefficient.

Keh and Ding (2000) derived the analytical relations for the case that the assumption of not overlapping EDLs can be removed. Ohshima (2019) derived the relations for a dilute suspension of spherical solid colloidal particles with a hydrodynamically slipping surface. However, weakening of the original Smoluchowski's assumptions naturally results in much more complicated relations.

In contrast to electroosmosis (Reuss 1809), the sedimentation potential was first observed approximately 70 years later by Dorn (1880). This is why the sedimentation potential is also referred to as the Dorn effect.

2.1.5 Characteristics of the total interaction energy curve

Based on the DLVO theory, the total interaction energy curve exhibits three important locations: primary minimum, energy barrier, and secondary minimum. It seems (Fig. 2.15) that the total energy permanently drops when approaching particle surface. However, due to the existence of strong repulsive forces at the tight vicinity of particle surface, the curve representing the total interaction energy attains its primary minimum. This minimum is accompanied by the presence of increased van der Waals attractive forces dominating the repulsive ones. The particles at these distances coagulate and their redistribution is improbable. With an increasing separation distance dominance of van der Waals forces over the EDL ones diminish, and the situation overturns to the opposite maximum called an energy barrier. To overcome this barrier (if exceeds the thermal energy) and to continue in coagulation process, it is necessary to exert maximum energy at this location. Consequently, the curve passes through its secondary minimum, where the attractive forces are higher than the repulsive ones. The coagulation process is again in progress, particles undergo weaker attraction in comparison with a location round the primary minimum, and as a result, the created flocs can be easily re-dispersed.

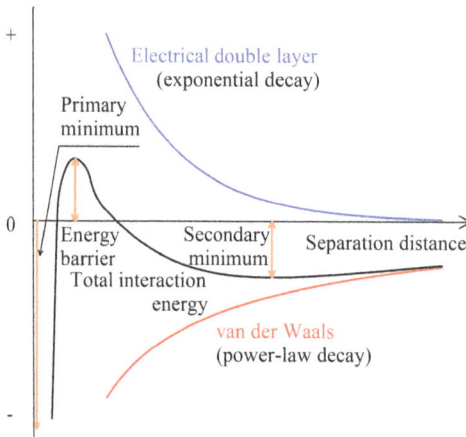

Fig. 2.15: Basic characteristics of the total interaction energy curve with an indication of both minima and maximal energy barrier.

For the given particles a value of the energy barrier strongly depends on temperature and pH values or their combination. An adequate choice of pH number can partially balance the problems with coagulation at lower temperatures. A significant dependence of energy barrier on pH values is illustrated in Fig. 2.16.

A decrease of energy barrier can be achieved with decreasing pH, appearance of surface roughness (in comparison with smooth surface, Hoek et al. 2003), decreasing the surface charge density, or increasing the ionic strength (Hernández 2023).

On the other hand, a decrease of barrier energy can result in a reduction of the secondary minimum as, for instance, in the case of small particles. Larger particles are accompanied by more progressive secondary minimum which contributes to the creation of more compact flocs (Hernández 2023).

The course of the total interaction energy was depicted (Fig. 2.15) for the case of the pure DLVO theory. The similar course also describes the XDLVO theory.

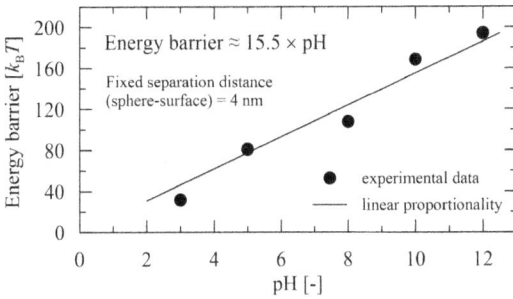

Fig. 2.16: A dependence of maximal energy barrier on pH (experimental data taken from Zhang et al. 2020; maximal energy barriers of anionic polyacrylamide (APAM) – polytetrafluoroethylene (PTFE) membrane).

2.1.6 Correctness of the ζ-potential evaluation

The preceding sections create an impression that the zeta potential is well-defined and its determination – for instance, by means of the traditional electrokinetic phenomena – is relatively straightforward. Unfortunately, the situation is much more complicated and every step concerning an introduction and consequent evaluation of the zeta potential raises a series of questions. The selected items are listed below.

The zeta potential is defined as a value which the potential curve attains at the location of shear (slipping) plane. The problem is that the location of this plane is unknown. It is obvious that this location is in a close proximity to the outer Helmholtz plane. However, even a location of this plane is unknown.

Apart from the notion 'zeta potential', the excess conductivity of the stagnant layer is another parameter describing behaviour of the EDL. Its significance is given by the dimensionless Dukhin number

$$Du = \frac{K^{surface}}{aK^{bulk}} \tag{2.26}$$

relating surface and bulk conductivities, where a is the local curvature radius of surface. The total conductivity K of the colloidal systems is the sum of the bulk conductivity and the weighted surface conductivity

$$K = K^{bulk} + \frac{Const}{a} \times K^{surface} = K^{bulk}(1 + Const \times Du), \tag{2.27}$$

where the value of Const subjects to surface geometry (Const = 2 for a cylindrical capillary). As the Helmholtz-Smoluchowski approach considers only bulk conduction, it is implicitly supposed $Du \ll 1$.

From the hydrodynamic viewpoint the shear plane separates the so-called hydrodynamically stagnant layer from the mobile one. This would imply that two physical characteristics, viscosity η and permittivity ε, should exhibit the discontinuous jumps at the shear plane which is not in accordance with the physical interpretation of the corresponding functional courses.

Andrade and Dodd (1951) proposed and experimentally verified the relation

$$\frac{d\eta(x)}{\eta(x)} = C_{VE} \times |E|^2, \tag{2.28}$$

where x is the distance from the surface and E is the electric field. This describes the so-called viscoelectric effect concerning an increase in viscosity of a polar liquid in an electric field. Lyklema and Overbeek (1961) reformulated the above relation to the form

$$\eta(x) = \eta_0 \left(1 + C_{VE} \times \left|\frac{d\psi}{dx}\right|^2\right), \tag{2.29}$$

where η_0 represents the constant value of bulk viscosity and $d\psi/dx$ is the double layer field strength. Hunter and Leyendekkers (1978) derived that the viscoelectric coefficient C_{VE} for water (a polar liquid due to the difference in electronegativities between oxygen and hydrogen atoms) attains a value within the interval $(0.5–1.0) \times 10^{-15}$ V^{-2} m^2. This illustrates rather complicated behaviour of a continuous distance-dependent viscosity in the tight proximity of surface. Behaviour of permittivity exhibits analogous features.

All these considerations can evoke a question whether the strict discrete definition of the shear plane could not be substituted by a term shear zone which could enable – among other things – gradual (however abrupt) changes of the physical parameters. For spherical particles this question is tightly connected with the values κa; in other words, with a ratio between an EDL thickness and a particle radius (see Fig. 2.17). For large values of κa, the diffuse layer thickness decreases, the ionic strength is intensive, and the potential gradient is apparent. In this case the values of surface, Stern and zeta potentials, are remarkably different. In the case of lower values of κa (the diffuse layer thickness dominates the particle radius) the strength is moderate, which gives a possibility to identify all three values.

If we detach the theoretical viewpoint, the similar problems can appear in the measurement of the zeta potential. It starts with the assumptions of smooth (up to molecular level) and homogeneous solid particles required by the models, and the problem can be also caused by appearance of impurities in the solutions or deformation of the particle surfaces.

It is not trivial to compare mutually the results (the values of zeta potential) based on the traditional electrokinetic phenomena. The experiments should be in compliance with the individual ranges of validity, the effects of relaxation and surface conductivity

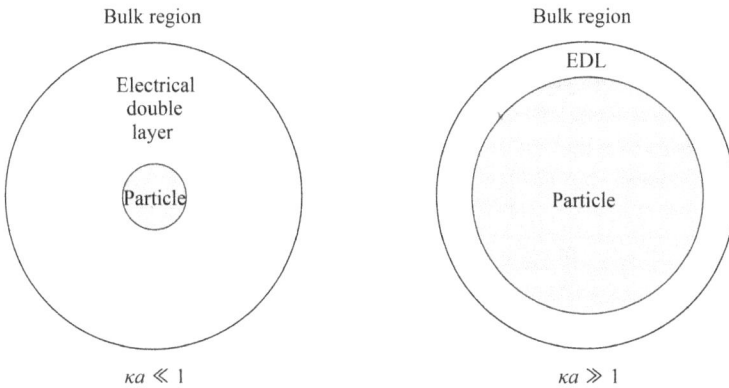

Fig. 2.17: Graphical interpretation of the κa relation.

should be negligible (as indicated above), and geometries of the EDLs should be similar as, for example, induced by a capillary used in electroosmosis and streaming potential methods. However, even under the fulfilment of all these assumptions the results from the individual experimental methods using various electrokinetic phenomena can differ. One explanation is presented in Szymczyk et al. (1998) who introduce that the location of the shear plane depends on the electrokinetic method used.

First attempts to compare the classical electrokinetic approaches appeared already in the 1930s in the contributions by Abramson and Grossman (1931) and Bull (1935). General feature in more recent contributions (Kim et al. 1996; Szymczyk et al. 1998) is obtaining the higher values of zeta potential by the electroosmosis method in comparison with the streaming potential one. A detailed comparison of both methods can be found in Plohl et al. (2020).

2.2 Schulze-Hardy and inverse Schulze-Hardy rules

An answer to the question how to enhance initiation and forming of aggregates is relatively simple: an application of so-called coagulants and – if necessary – the appropriate additives. However, this answer is accompanied by a variety of assorted aspects: material availability, solubility, minimal dosing, financial costs, and efficiency – to name just the basic ones.

2.2.1 Classical Schulze-Hardy rule

The Schulze-Hardy rule addresses the last aspect: efficiency. However, a derivation of the Schulze-Hardy rule indicating the attributes of the coagulants to choose is not so straightforward and faces the shortcomings of the DLVO theory when applied to water treatment. A derivation of the Schulze-Hardy rule, as initially experimentally deduced by Schulze (1882) and Hardy (1900) and later on supported by the DLVO balance equation, comes out from two principal assumptions: (a) an application of symmetric electrolytes and (b) a high level of surface potential. To these points it is also necessary to add an assumption of planar surfaces of unwanted impurities. Based on these assumptions, the Schulze-Hardy rule relates the critical coagulation concentration (CCC) c_{cr} (the minimum concentration of counterions to induce instability of a colloidal dispersion) and the valence z of these counterions

$$c_{cr} = \text{Const} \times \frac{(\tanh(zy^d))^4}{A^2 z^6},$$

(2.30)

where y^d is the dimensionless potential of the diffuse part of the EDL

$$y^d = \frac{e}{k_B T} \times \psi^d,$$

(2.31)

where e is the elementary charge ($=1.602176634 \times 10^{-19}$ C), k_B is the Boltzmann constant ($=1.380649 \times 10^{-23}$ J·K^{-1}), $k_B T$ represents the unit of thermal energy, T is the absolute temperature, and A is the Hamaker constant.

Intuitively, the Schulze-Hardy rule relates a CCC with the −6 power of counterion valency, thus predicting relatively steep decrease of CCC with an increasing valency. For instance, for $z = 1,2,3$ we obtain $c_{cr} \sim 1/1, 1/64, 1/729$. Multiplying these fractions by 729 we get the ratio between the impacts of the consequent valencies as 729:11.4:1. This indicates that the amount of electrolytes (dissociating into ions in a carrier liquid and thus conducting electricity) very strongly depends on the counterion valencies, and an amount of electrolyte 3:3 can be 729 times lower than that for electrolyte 1:1 with the same impact on destabilising of colloidal system.

It is necessary to keep in mind that the electrolytes are supposed to be symmetric (all ions are of comparable charge and size) as well as a relatively high value of surface potential. Such value enables a replacement of the nominator (tanh) in rel. (2.30) by its limiting value 1. This is fulfilled if the argument in tanh attains approximately a value 5, $\tanh(5) \approx 1$ (see Fig. 2.18). For a temperature $T = 293.15$ K, the normalised coefficient $e/(k_B T)$ attains the approximate value 39.6. From here it follows that the values of a diffuse potential should exceed $\psi^d = 126$ mV for monovalent counterions $z = 1$, $\psi^d = 63$ mV for bivalent counterions $z = 2$, and $\psi^d = 42$ mV for trivalent counterions $z = 3$. For lower values of ψ^d (when the limiting value 1 of tanh is not attained) it is necessary to express (exactly as tanh is an analytic function) a function tanh by means of its Taylor series

$$\tanh(x) = x - \frac{x^3}{3} + \frac{2x^5}{15} - \frac{17x^7}{315} + \dots ; |x| < \frac{\pi}{2}$$

(2.32)

Hence, for lower values of the diffuse potential the function tanh can be substituted by its argument

$$c_{cr} = \text{Const} \times \frac{\left(zy^d\right)^4}{A^2 z^6}, \tag{2.33}$$

which results in the relation

$$c_{cr} = \text{Const} \times \frac{\left(y^d\right)^4}{A^2} \times z^{-2}. \tag{2.34}$$

It substantially changes the dependence of a CCC on valency of counterions of the applied electrolyte. The decay of CCC is rather stronger than inverse quadratic due to a decrease of the diffuse potential ψ^d with valency z (Lyklema 2013).

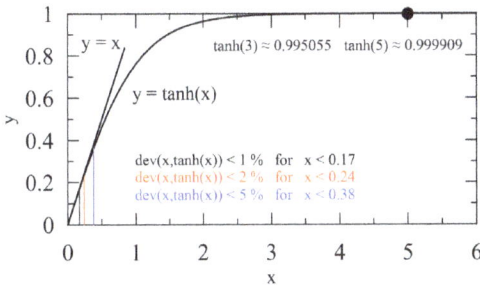

Fig. 2.18: A course of the function tan*h* with the deviations from its argument for the specific values.

2.2.2 Modified Schulze-Hardy rule

As already indicated above, practically all the assumptions applied above are in contradiction with the requirements in some industrial branches such as water treatment where aggregation of undesired particles is a key topic. Based on numerous experiments, a qualitative validity of the Schulze-Hardy rule is doubtless: a decrease of CCC with increasing value of counterion valency, the question is at which rate.

The assumption of symmetric electrolytes can be questioned in the following points:

– multivalent symmetric z:z electrolytes are usually hardly soluble (especially in water), thus preventing the dissociated ions from relatively uniform distribution within a solvent containing particles;
– symmetric electrolytes (comprised of ions with similar charges and sizes) exhibit equal mobility of ions and uniform conductivity (i.e. electricity is conducted through the movement of ions, not through the movement of electrons), this contradicts unpredictable behaviour of ions dissociated from the asymmetric electrolytes.

The assumption of a high level of diffuse potential cannot be met in practice and is unrealistic. These factors stimulated an orientation to more realistic (in water treatment) asymmetric 1:z and z:1 electrolytes and also to a replacement of two parallel plates by other geometries as for instance by spherical shapes of the particles (Hsu et Kuo 1977; Hsu and Liu 1998; Chu et al. 2018).

A better insight into this problem can be provided by the more detail description of a course of potential in the EDL. To this aim there exist two equations relating potential with entry parameters, specifically the Poisson-Boltzmann equation and its linearised (hence much simpler) form the Debye-Hückel equation already introduced in eqs. (1.42) and (1.43). The complete (non-linearised) Poisson-Boltzmann equation valid between two charged plates is of the form (the origin of a longitudinal distance x is located in the mid-plane; see Fig. 1.16):

$$\frac{d^2\psi}{dx^2} = -\frac{e}{\varepsilon_r \varepsilon_0} \sum_i c_i z_i \cdot e^{-\frac{ez_i}{k_B T}\psi}, \tag{2.35}$$

where ε_r is the relative permittivity of the solution, ε_0 is the permittivity of a vacuum ($=8.8541878188(14) \times 10^{-12}$ F·m^{-1}), c_i is the molar concentration of ions of the type i, and z_i is their corresponding charge expressed in units of the elementary charge e ($=1.602176634 \times 10^{-19}$ C). The summation is carried out over all the types of ions. This equation can be solved numerically, the analytical solutions known only for a couple of special cases (for instance, symmetric $z{:}z$ and asymmetric $z{:}2z$ electrolytes) are based on so-called special functions, dominantly either on Jacobi or Weierstrass elliptic functions (the term elliptic is taken over from the elliptic integrals by means of which it is possible to determine a length of the elliptic arcs) (see e.g. Behrens and Borkovec 1999; Polat and Polat 2010; Xing 2011; Johannessen 2012; Zhang et al. 2018; Šamaj and Trizac 2019; Travesset 2024). A detailed approach to the analysis of the Poisson-Boltzmann equation is presented in Trefalt et al. (2013).

An example of the Poisson-Boltzmann equation exemplary demonstrates that a solution of any equation does not subject only to a functional complexity of the equation itself but is equally interlaced with the form (geometry) and functional complexity of the boundary (and/or initial) conditions. In this case it means feasibility of a solution with a change of particle shape (for instance, a passage from smooth plates to a spherical shape) or redistribution of surface charges. It implies that under a preservation of the same (balance) equation a passage to other boundary (initial) conditions represents a completely new separate problem (see also Trefalt et al. 2015).

In contrast to the Poisson-Boltzmann equation applicable for higher values of surface potential, its linearised (only a linear term – with the negative sign – in the Taylor expansion is considered) Debye-Hückel (DH) form is valid for lower values of surface potential and in far-field regime

$$\frac{d^2\psi(x)}{dx^2} = \frac{e^2\Sigma c_i z_i^2}{\varepsilon k_B T}\psi(x).\tag{2.36}$$

The linearised DH description is applicable for potential magnitudes up to ~25 mV for symmetric 1:1 electrolytes at ambient temperature. On the other hand – in the presence of multivalent ions – it can fail even at much lower values of potential (Trefalt et al. 2013).

Following Trefalt et al. (2013) the DH equation can be rewritten into the form

$$\frac{d^2\psi(x)}{dx^2} = \frac{2e^2 I}{\varepsilon k_B T}\psi(x)\tag{2.37}$$

introducing the ionic strength

$$I = \frac{1}{2}\sum_i c_i z_i^2.\tag{2.38}$$

Indisputable advantage of the linearised form is its much better computational handling. This enables to derive a relation for the case of asymmetric 1:z or z:1 electrolytes (Trefalt et al. 2013)

$$I = \frac{z(z+1)}{2}c.\tag{2.39}$$

However, the explicit relations between a CCC and contraion valency are still unclear.

If we consider electrolyte mixture composed of two electrolytes with the molar concentrations c_A and c_B and the corresponding individual CCCs $c_{cr,A}$ and $c_{cr,B}$, then Grolimund et al. (2001) proposed the following relation evaluating CCC of electrolyte mixture on the basis of experimental measurements:

$$\frac{1}{c_{cr}} = \frac{1-q}{c_{cr,A}} + \frac{q}{c_{cr,B}},\tag{2.40}$$

where the quotient q relates c_B to the total electrolyte concentration ($=c_A + c_B$). Later this expression was theoretically derived by Trefalt et al. (2013) for a mixture of 1:1 and 1:z electrolytes.

2.2.3 Inverse Schulze-Hardy rule

Starting from the pioneering works of Schulze (1882) and Hardy (1900), the Schulze-Hardy rule is traditionally based on a relation between CCC and the valency of counterions. Cao et al. (2015) were the first who studied the dependence between CCC and multivalent coions having the same charge as colloidal particles. Analogously to the classical Schulze-Hardy rule (power-law relation) they formulated the relation (later theoretically derived by Trefalt (2016))

$$c_{cr} \propto z^{-1}, \tag{2.41}$$

where – this time – z represents the valency of multivalent coions. They denoted this relation as the inverse Schulze-Hardy rule.

As apparent the dependence on the valence is (much) weaker than in the case of the valency of multivalent counterions. Behaviour of asymmetric z:1 or 1:z electrolytes significantly differs. The multivalent counterions with the opposite sign of charge than the colloidal particles adsorb to the particle surface thus remarkably reducing the magnitude of the surface charge. The multivalent coions with the same sign of charge as the colloidal particles are repelled from the particles which results in preserving the magnitude of the surface charge (Trefalt et al. 2017).

In comparison with the classical Schulze-Hardy rule a number of experimental data for verification of the inverse Schulze-Hardy case is scarce (Trefalt et al. 2020).

2.3 Coagulants – basic categorisation

Two consequent processes – coagulation and flocculation – can be, in spite of their partial overlapping, relatively well determined as for the first time strictly distinguished by La Mer and Healy (1963). During the first relatively short phase (lasting about 10 s) denoted as coagulation the principal aim is to destabilise a given colloidal suspension or solution. During the second phase (lasting in the tens of minutes) – flocculation – the destabilised particles should aggregate which creates the condition for efficient settling and/or filtering.

Terminology using the words coagulant and flocculant cannot be so strict as primarily used coagulants are also present in the phase of flocculation. Nevertheless, the flocculants (often organic polyelectrolytes) are added with the onset of flocculation.

The long-term used traditional coagulants reducing the repulsive potential of EDL are aluminium and iron salts supplying the counterions to tight proximity (a diffuse layer) of (usually) negatively charged particle surface. The EDL is compressed and the particles can exceed the van der Waals contact distance which results in forming of aggregates. With respect to the Schulze-Hardy rule the choice of these trivalent metals (Fe^{3+} and Al^{3+}) is not surprising and the inorganic salts such as $AlCl_3$, $Al_2(SO_4)_3$, $FeCl_3$, or $Fe_2(SO_4)_3$ are used. However, the classical coagulant agents exhibit a series of shortcomings as necessary pH adjustment or higher dosing resulting among other in excessive sludge production. Moreover, aluminium is far from optimal what concerns health approach. On the other hand, availability of Fe and Al salts is quite easy and they are relatively cheap.

With the onset of the twenty-first century the classical (Al, Fe) approach was forked to more directions:
-   The experiments showed better performance of tetravalent (cf. Schulze-Hardy rule) metals such as Ti (Zhao et al. 2011; Hussain et al. 2014; Hussain et al. 2019;

Gan et al. 2021) or Zr (Jarvis et al. 2012; Hussain et al. 2014), among other things creation of larger and stronger flocs or a conversion of Ti coagulation sludge into high value-added products (TiO_2 photocatalysts). However, it is also necessary to consider health hazard and financial expenses.

– Better efficiency than the traditional salts was also achieved by pre-polymerisation of metal-ion coagulants. As a result, it can be mentioned amply used polyaluminium chloride (PAC or PACl), polyaluminium sulfates, polyaluminium chlorosulfates, and polyferric sulfate. Their application is connected with lower dosing, looser pH choice, and lower sensitiveness to temperature changes.

– Another trend is connected with an application of tetravalent (cf. Schulze-Hardy rule) metals as modifiers of the existing coagulants such as silicon (Si^{4+}) in polyferric silicate chloride and polyaluminium silicate chloride (Tzoupanos and Zouboulis 2008; Tzoupanos et al. 2009), or silicon (Si^{4+}) and zirconium (Zr^{4+}) in polyaluminium zirconium silicate coagulant (Zhuang et al. 2021). A combination of Al, Fe, and Si was analysed in Gao et al. (2006). The introductory steps are also carried out with the hexavalent ferrate; however, its potential application is accompanied by poor stability and difficult storage (De Marines et al. 2024).

– With permanent deterioration of raw water supplying the water treatment plants and consequent measures contributing to improvement of the environment there is a need to meet this problem with an adequate choice of coagulants substantially reducing the impact on the environmental pollution. In this context metal coagulants produce large volume of hazardous sludges in contrast to the biocoagulants which are biodegradable. On the other hand, efficiency and the length of retention time are (much) more favourable for the metal coagulants. The aim is to prepare such combinations which optimise these parameters. The source for extracting biocoagulants is, for instance, crustaceans (a linear polysaccharide chitosan – Zeng et al. 2008; Yang et al. 2016) and plant seeds (*Moringa oleifera* – Yamaguchi et al. 2021). Detailed lists of various biocoagulants/bioflocculants are listed, e.g., in Saleem and Bachmann 2019; Maćczak et al. 2020; Jiang et al. 2021; Okoro et al. 2021; Kurniawan et al. 2022).

The Schulze-Hardy rule indicates a relation between valency of the used metals and its potential efficiency. Nevertheless, already in 1928, Mattson (1928) showed that the (rapid) hydrolysis products of Al and Fe salts were more important than their trivalent ions. The products significantly participate in the electrical neutralisation (Duan and Gregory 2003; Gregory and Duan 2001).

Detail information on coagulants can be found, for instance, in Bolto and Gregory (2007), Jiang (2015), and Dayarathne et al. (2021). This will be also partially the topic of the following chapter.

Proper choice of coagulants is unavoidably interlaced with a determination of suitable pH range (as will be discussed in Chapter 3) and an adequate dosage. Gregory and Duan (2001) classified dosing into four zones:

- Zone 1: Very low coagulant dosage; particles still negative and hence stable;
- Zone 2: Dosage sufficient to give charge neutralisation and hence coagulation;
- Zone 3: Higher dosage giving charge neutralisation and restabilisation;
- Zone 4: Still higher dosage giving hydroxide precipitate and sweep flocculation.

The possibilities how to evaluate an adequate dosage are presented in the following section.

2.4 Coagulant dosage

Loosely speaking, there are three crucial factors participating in successful water treatment: a choice of a coagulant agent, its dosage, and a value of pH. In the following three evaluation methods concerning a dosage will be presented, and other methods can be found in the historical follow-up reviews (TeKippe and Ham 1970a,b; Dentel 1991; Ratnaweera and Fettig 2015) and the references in the cited literature. At the beginning the dominant and indispensable jar test will be presented followed by artificial networks techniques and classical regression methods.

2.4.1 Jar test

As apparent from Chapter 1 and the preceding sections in this chapter, a determination of the individual parameters indicating a tendency to destabilisation of the colloid systems may be rather inaccurate. This has a series of reasons. The thorniest problem is in setting a balance equation. The classical DLVO approach fails in the case of aqueous solutions, an introduction of the Lewis acid-basic term in the balancing relations substantially improves the results; however, the whole situation is more delicate.

To make the water treatment process more efficient, the agents called coagulants have to be used to accelerate efficiency of destabilisation and to enhance formation of initial particle clusters (flocs). A proper choice of coagulants subjects to quality and composition of treated water and its temperature. Based on the input characteristics, an optimising procedure has to be carried out: a choice of an adequate coagulant, its amount, and a value of pH under which is applied (consequent geometrical and hydromechanical arrangements in the water treatment plants will be presented in Chapter 3).

The aim is to minimise a coagulant dosage not only from the viewpoint of cost effectiveness. Under-dosing results in insufficient removal of pollutants, and overdosing (in an extreme case it can restructure stability of the suspensions) may evoke health problems and worsens processing of so-called sludge (waste products during water treatment).

All these arguments justify a common usage of the jar test as a traditional method for an evaluation of three basic inputs: suitable choice of coagulant, an adjustment of an appropriate value of pH, and an adequate dosage. This test simulates – to a certain measure – mixing and settling conditions in a clarifier. The jar test equipment (see Fig. 2.19) is a stirring device composed usually of four to eight jars (beakers) of equal volume (containing 0.5, 1, or 2 L of raw water). Each baffle-free jar is equipped with centrally placed shaft holding a horizontally or vertically placed paddle(s). Nevertheless, various geometrical arrangements of the impellers are used.

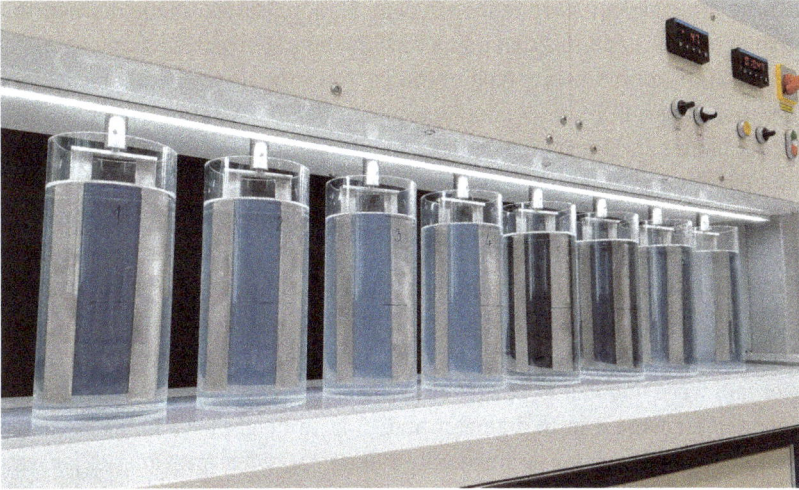

Fig. 2.19: A jar test device equipped with eight 2 L jars agitated with vertically placed twin paddles.

A jar test experiment comprises three consequent steps: pre-homogenisation of a coagulation agent, a short-term coagulation stage (relatively rapid mixing in seconds), and finally a long-term flocculation stage (in minutes applying much lower mixing intensity minimising flocs breakage). The principle aim is to evaluate floc size evolution which is based on two procedures:
– a suitable illumination of jar contents which can be achieved, for instance, by using a particle image velocimetry (PIV) non-invasive technique creating a two-dimensional sheet in combination with a high-speed digital camera perpendicularly placed with respect to a laser source (see Bouyer et al. 2001; Xiao et al. 2011);
– it is necessary to process digital records via a digital image technique (more detailed information is presented in Chapter 4).

The advantage of the PIV technique over the classical photographic one is in eliminating – to a certain degree – uncertainties caused by superposable flocs and – due to intensive light sheet – capturing more accurate and bright floc pictures. However, the classical problems with a projection of non-spherical particles still persist.

Common usage of the jar test can be improved by simultaneous application of another technique providing more parallel pieces of information. Cheng et al. (2008) and Cheng et al. (2011) implemented a nephelometric turbidimeter through a hole in the beaker using a light-emitting diode (LED). Such arrangement enables a simultaneous recording of two basic parameters in the process of water treatment: sedimentation rate and residual turbidity after a period of settling. A degree of the observed turbidity is also interlaced with a determination of coagulant dosing. Simultaneous measurement of turbidity was also carried out by Tassinari et al. (2015) who added six turbidimeters coupled with a data-acquisition system to a commercially produced jar tester. All these arrangements are more efficient than a classical method when the samples have to be taken and inserted into a turbidimeter or spectrophotometer.

Fujisaki (2018) proposed a new type of the jar tester by attaching photocouplers (devices incorporating a LED and a photodetector in one package) to both sides of a beaker. During pauses in intermittent mixing the floc settling velocity and the residual solution turbidity were measured by light transmission.

Summarising more than a 100-year preceding history of an application of jar tests in drinking water treatment, there is a possibilily to identify one crucial attribute: lack of any standardisation. First pre-jar tests appeared already at the turn of the twentieth century (in more detail in Black et al. 1957). However, the first publication on jar tester composed of four jars (see Fig. 2.20) appeared in 1921 by Langelier (1921). A number of jars were doubled in Langelier and Ludwig (1949). While in 1921 the dimensions of each jar were 5 in in diameter and 10 in in height (~3.2 L), in the latter publication a volume of each jar attained only 200 mL. Massive spreading of jar testing was not (has not been) accompanied by its uniformity.

Fig. 2.20: Langelier's jar tester (taken from Black et al. 1957, Fig. 1).

Already in 1957 Black et al. (1957) – 36 years after the first publication on jar testing –
addressed the laboratories with the demand to adhere to these five points:
– the size of the sample,
– the size and shape of the container,
– peripheral speed and time of rapid mixing,
– peripheral speed and time of slow mixing,
– criteria for establishing optimum dosage.

It is possible to say that at present the situation is identical with that before many
decades. It really limits any specific comparison of the individual results and some-
how prevents the vast range of the worldwide measured data from their mutual detail processing.

The principal problem consists in the fact that the identical velocity gradients
(usually denoted as G) can be achieved with different geometries of jars and impellers. However, the corresponding flow fields are completely different in local stress
distributions. This also generates a completely different particle size distribution.

2.4.2 Artificial intelligence techniques

Absence of uniformity projects to a newly developing trend in optimising coagulant
dosage: an application of various artificial intelligence (AI) techniques based on jar-
test data. Basically these techniques are carried out in two steps: a sufficiently wide
set of empirical data as an input and a choice of an adequate computational technique. Due to non-uniformity of the experimental setups and experimental data processing, the 'feeding' in the first step is far from being optimal. On the other hand,
there is a series of AI techniques as presented in Fig. 2.21 using a list introduced for
water treatment in Fan et al. (2018).

Biological neural networks with their synapses in human brains inspired a creation of computer-based neural networks models processing a set of input data (in the
so-called input layer) through a cascade of hidden layers (denoted as shallow in the
case of only one hidden layer and deep for two and more, see Figs. 2.22 and 2.23). Each
node in any hidden layer is supplied by information from the nodes in the preceding
layer (indicated by the edges) through a so-called activation function. This function
represents non-linear processing of the inputs in the preceding layer. The processed
pieces of information are advanced to the consequent layer, and finally, the outcome
is provided by the nodes in the last – output – layer. The quality of the non-linear
transformations corresponds to the quality of their training, i.e. it is based on the results carried out using real jar testers and their classical evaluation. The more diverse
and numerous are these sets or real data, the more elaborated are the 'trained' transformations (mathematical relations). In water treatment the input layer is composed
of pH values as a key parameter accompanied by other parameters such as turbidity,

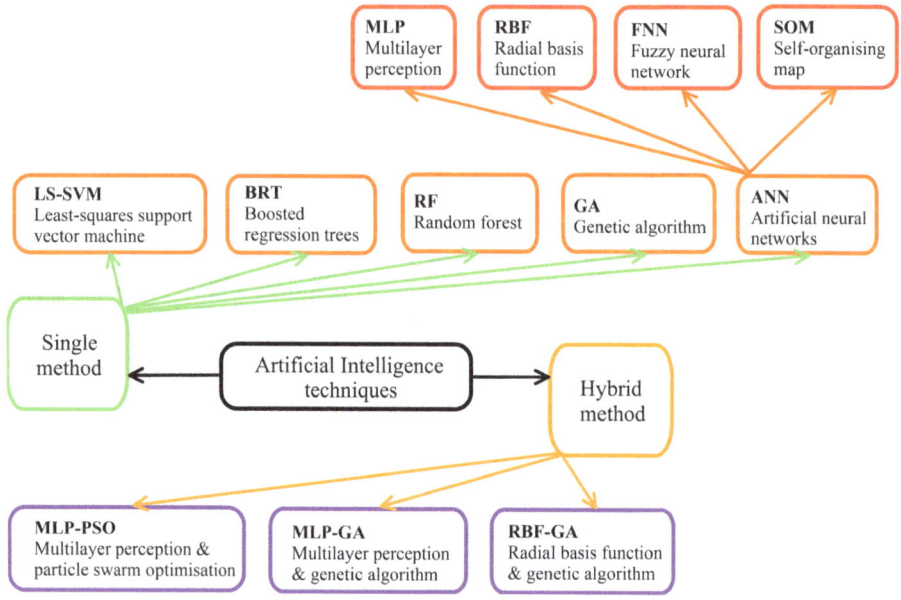

Fig. 2.21: AI techniques used in water treatment.

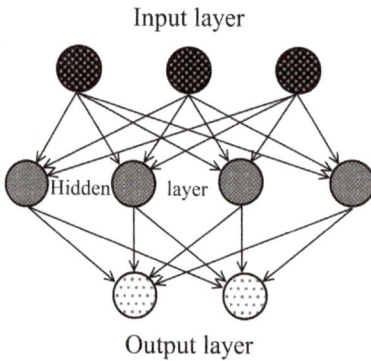

Fig. 2.22: The shallow neural network.

alkalinity, temperature, electrical conductivity, suspended solids, UV-absorbance, colour, and total organic carbon (TOC – indicating water contamination by synthetic organic compounds). The output layer is usually formed by the only node representing coagulant dosage.

An application of AI techniques to an evaluation of coagulant dosage in water treatment started more intensively approximately at the turn of the twenty-first century (Wu and Lo 2008, and others) with a general access to the advanced relatively inexpensive computational technique. The hitherto achieved results are summarised, for instance, in Li et al. (2021), Alam et al. (2022), and Liu et al. (2024).

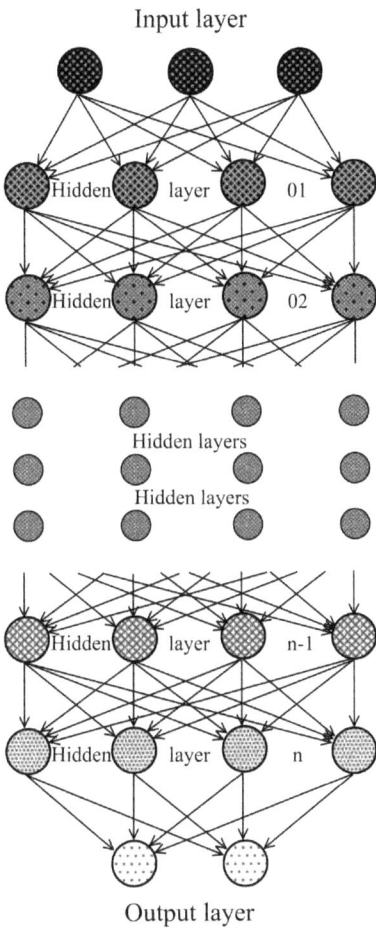

Fig. 2.23: The deep neural network.

If the traditional jar tests and AI approaches are compared, there are advantages and disadvantages on both sides. A jar test is a reliable method for the determination of a coagulant dosage. However, it is time-consuming which can cause a time delay in dosage optimisation with abruptly changing attributes of processed water as, for instance, pollutant density during heavy rain. It also requires skilful staff and laboratory equipment. Nevertheless, the last shortcomings are only illusive as insufficient historical jar-test data can cause severe errors in AI predictions. The shortcoming concerning time delay and reaction to abrupt changes can be also valid for AI techniques if not sufficiently trained in these situations. The problems can also occur if the AI procedures are transformed to other water treatment plants, where they were not trained. Moreover, efficiency of any method used in the water treatment plants is expected to be 100%; otherwise it cannot be applied. In spite of all these discussible items, development of the AI techniques seems to be very promising (and saving production costs in practice), and simultaneously it should be repeated that standardisation of a jar-test procedure should accelerate a progress in this trend.

It is necessary to mention that AI techniques have no relation to the physical and/or chemical modelling and are purely based on mathematical evaluation and processing of sufficiently large data sets optimally covering all four seasons and acquired repeatedly in consequent years.

Another question is what is hidden behind a term 'optimisation of coagulant dosage'. Jar testing itself is only an approximation of real conditions in the water treatment plants differing both in geometrical arrangement and also in flow conditions. In this sense the word 'optimisation' should be replaced by the word 'approximation'. It raises a question whether at the concrete plants the training data sets in the AI techniques should not be also enlarged by real relations: applied dosage vs. other input parameters.

2.4.3 Regression analysis

Adequate coagulant dosage can be also approximated using regression models, which are usually of linear nature. This does not mean that the functional forms of entry parameters have to exhibit linearity but can appear, e.g. as $Temp^2$ or $\log(Temp)$, where $Temp$ denotes temperature. The term linear is related to the corresponding coefficients with these parameter functions as Jackson and Tomlinson (1986):

$$\text{Iron sulfate dose (mg/L)} = C_0 + C_1 \times Turb + C_2 \times \log(Turb) + C_3 \times Cond + C_4 \times \log(Cond) +$$

$$C_5 \times Temp + C_6 \times \log(Temp) + C_7 \times Flow + C_8 \times \log(Flow). \tag{2.42}$$

Here all the coefficients C_i ($i = 1, \ldots, 8$) appear in the linear form. The term non-linear is connected with an occurrence of the parameters as variables in the non-linear functions (for instance, in the power form $(C_i)^a$, $a \neq 1$).

The proposed models of this type are relatively simple but can be efficient in the individual water treatment plants especially when the initial set of data covers sufficiently long time period. The resulting values of the coefficients also indicate the input weights among the individual entry parameters.

More complicated is the multivariate regression where a number of input and output parameters are at least two in both cases. Not always such approach provides a satisfactory conclusion as demonstrated by Franceschi et al. (2002) who analysed experimental responses of residual turbidity and minimal coagulant dosage. They received that a fulfilment of both requirements (low residual turbidity and minimal coagulant dosage) is in contradiction. Hence, in this case a criterion on low residual turbidity was preferred.

References

Abramson, H.A., Grossman, E.B. (1931) Electrokinetic phenomena: IV. A comparison of electrophoretic and streaming potentials. Journal of General Physiology 14, 563–573. https://doi.org/10.1085/jgp.14.5.563

Alam, G., Ihsanullah, I., Naushad, M., Sillanpää, M. (2022) Applications of artificial intelligence in water treatment for optimization and automation of adsorption processes: Recent advances and prospects. Chemical Engineering Journal 427, 130011. https://doi.org/10.1016/j.cej.2021.130011

Andrade, E.N. da C., Dodd, C. (1951) The effect of an electric field on the viscosity of liquids. II. Proc. Roy. Soc. (London) A204, 449–464. https://doi.org/10.1098/rspa.1951.0002

Behrens, S.H., Borkovec, M. (1999) Exact Poisson-Boltzmann solution for the interaction of dissimilar charge-regulating surfaces. Physical Review E 60, 7040–7048. https://doi.org/10.1103/PhysRevE.60.7040

Black, A.P., Buswell, A.M., Eidsness, F.A., Black, A.L. (1957) Review of the Jar Test. Journal of the American Water Works Association 49, 1414–1424. https://www.jstor.org/stable/41254753

Bolto, B., Gregory, J. (2007) Organic polyelectrolytes in water treatment. Water Research 41, 2301–2324. https://doi.org/10.1016/j.watres.2007.03.012

Booth, F. (1954) Sedimentation potential and velocity of solid spherical particles. Journal of Chemical Physics 22, 1956–1968. https://doi.org/10.1063/1.1739975

Bouyer, D., Line, A., Cockx, A., Do-Quang, Z. (2001) Experimental analysis of floc size distribution and hydrodynamics in a jar-test. Chemical Engineering Research & Design 79, 1017–1024. https://doi.org/10.1205/02638760152721587

Bull, H.B. (1935) Electrokinetics. XIV A critical comparison of electrophoresis, streaming potential, and electroosmosis. Journal of Physical Chemistry 39, 577–583. https://doi.org/10.1021/j150365a001

Camp, T.R., Root, D.A., Bhoota, B.V. (1940) Effects of temperature on rate of floc formation. Journal American Water Works Association 32, 1913–1927. https://doi.org/10.1002/j.1551-8833.1940.tb19608.x

Cao, T., Szilágyi, I., Oncsik, T., Borkovec, M., Trefalt, G. (2015) Aggregation of colloidal particles in the presence of multivalent co-ions: The inverse Schulze–Hardy rule. Langmuir 31, 6610–6614. https://doi.org/10.1021/acs.langmuir.5b01649

Chapman, D.L. (1913) A contribution to the theory of electrocapillarity. The London, Edinburgh, and Dublin Philosophical Magazine and Journal of Science 25, 475–481. https://doi.org/10.1080/14786440408634187

Cheng, W.P., Chen, P.H., Yu, R.F., Hsieh, Y.J., Huang, Y.W. (2011) Comparing floc strength using a turbidimeter. International Journal of Mineral Processing 100, 142–148. https://doi.org/10.1016/j.minpro.2011.05.010

Cheng, W.P., Kao, Y.P., Yu, R.F. (2008) A novel method for on-line evaluation of floc size in coagulation process. Water Research 42, 2691–2697. https://doi.org/10.1016/j.watres.2008.01.032

Chu, Y., Chen, J., Haso, F., Gao, Y., Szymanowski, J.E.S., Burns, P.C., Liu, T. (2018) Expanding the Schulze–Hardy rule and the Hofmeister series to nanometer-scaled hydrophilic macroions. Chemistry – A European Journal 24, 5479–5483. https://doi.org/10.1002/chem.201706101

Dayarathne, H.N.P., Angove, M.J., Aryal, R., Abuel-Naga, H., Mainali, B. (2021) Removal of natural organic matter from source water: Review on coagulants, dual coagulation, alternative coagulants, and mechanisms. Journal of Water Process Engineering 40, 101820. https://doi.org/10.1016/j.jwpe.2020.101820

Dayarathne, H.N.P., Angove, M.J., Jeong, S., Aryal, R., Paudel, S.R., Mainali, B. (2022) Effect of temperature on turbidity removal by coagulation: Sludge recirculation for rapid settling. Journal of Water Process Engineering 46, 102559. https://doi.org/10.1016/j.jwpe.2022.102559

De Marines, F., Corsino, S.F., Castiglione, M., Capodici, M., Torregrossa, M., Viviani, G. (2024) Ferrate as a sustainable and effective solution to cope with drinking water treatment plants challenges. Journal of Environmental Chemical Engineering 12, 112884. https://doi.org/10.1016/j.jece.2024.112884

Debye, P., Hückel, E. (1924) Bemerkungen zu einem Satze die kataphoretische Wanderungs geschwindichkeit suspendierter Teilchen. Physikalische Zeitschrift 25, 49–52.

Delgado, A.V., González-Caballero, F., Hunter, R.J., Koopal, L.K., Lyklema, J. (2007) Measurement and interpretation of electrokinetic phenomena. Journal of Colloid and Interface Science 309, 194–224. https://doi.org/10.1016/j.jcis.2006.12.075

Dentel, S.K. (1991) Coagulant Control in Water Treatment. Critical Reviews in Environmental Control 21, 41–135. https://doi.org/10.1080/10643389109388409

Dorn, E. (1880) Ueber die Fortführung der Electricität durch strömendes Wasser in Röhren und verwandte Erscheinungen. Annalen der Physik und Chemie 10, 46–77. https://www.google.com/url?sa= t&source=web&rct=j&opi=89978449&url=https://scholar.archive.org/work/bqtsxgf3srcjbpgyk pi36hw5za/access/ia_file/crossref-pre-1909-scholarly-works/10.1002%25252Fandp.18772360313.zip/ 10.1002%25252Fandp.18802460505.pdf&ved=2ahUKEwjW0obWh4eLAxUa0QIHHfrmNEIQFnoECBkQA Q&usg=AOvVaw2lFq_YzbNfSKXhTTzaN3H_

Duan, J., Gregory, J. (2003) Coagulation by hydrolysing metal salts. Advances in Colloid and Interface Science 100–102, 475–502. https://doi.org/10.1016/S0001-8686(02)00067-2

Elimelech, M., Gregory, J., Jia, X., Williams, R.A. (1995) Particle Deposition and Aggregation: Measurement, Modelling and Simulation. Butterworth-Heinemann.

Ellison, W.J. (2007) Permittivity of pure water, at standard atmospheric pressure, over the frequency range 0 – 25 THz and the temperature range 0 – 100 °C. Journal of Physical and Chemical Reference Data 36, 1–18. https://doi.org/10.1063/1.2360986

Fan, M.Y., Hu, J.W., Cao, R.S., Ruan, W.Q., Wei, X.H. (2018) A review on experimental design for pollutants removal in water treatment with the aid of artificial intelligence. Chemosphere 200, 330–343. https://doi.org/10.1016/j.chemosphere.2018.02.111

Franceschi, M., Girou, A., Carro-Diaz, A.M., Maurette, M.T., Puech-Costes, E. (2002) Optimisation of the coagulation-flocculation process of raw water by optimal design method. Water Research 36, 3561–3572. https://doi.org/10.1016/s0043-1354(02)00066-0

Fujisaki, K. (2018) Experimental study on flocculation performance of chitosan-based flocculant using a novel jar tester. Journal of Civil Engineering and Environmental Sciences 4, 38–43. http://doi.org/10. 17352/2455-488X.000026

Gan, Y., Li, J., Zhang, L., Wu, B., Huang, W., Li, H., Zhang, S. (2021) Potential of titanium coagulants for water and wastewater treatment: Current status and future perspectives. Chemical Engineering Journal 406, 126837. https://doi.org/10.1016/j.cej.2020.126837

Gao, B.-Y., Yue, Q.-Y., Wang, B.-J. (2006) Properties and coagulation performance of coagulant poly-aluminum-ferric-silicate-chloride in water and wastewater treatment. Journal of Environmental Science and Health, Part A 41, 1281–1292. https://doi.org/10.1080/10934520600656869

Gouy, G (1910) Sur la constitution de la charge électrique à la surface d'un électrolyte. Journal de Physique Théorique et Appliquée 9, 457–468. https://doi.org/10.1051/jphystap:019100090045700

Grahame, D.C. (1947) The electrical double layer and the theory of electrocapillarity. Chemical Reviews 41, 441–501. https://doi.org/10.1021/cr60130a002

Gregory, J., Duan, J. (2001) Hydrolyzing metal salts as coagulants. Pure and Applied Chemistry 73, 2017–2026. https://doi.org/10.1351/pac200173122017

Grolimund, D., Elimelech, M., Borkovec, M. (2001) Aggregation and deposition kinetics of mobile colloidal particles in natural porous media. Colloids and Surfaces A: Physicochemical and Engineering Aspects 191, 179–188. https://doi.org/10.1016/S0927-7757(01)00773-7

Haarhoff, J., Cleasby, J.L. (1988) Comparing aluminum and iron coagulants for in-line filtration of cold water. Journal of the American Water Works Association 80, 168–175. https://doi.org/10.1002/j.1551-8833.1988.tb03022.x

Hanson, A.T., Cleasby, J.L. (1990) The effects of temperature on turbulent flocculation: Fluid dynamics and chemistry. Journal of the American Water Works Association 82, 56–73. https://doi.org/10.1002/j.1551-8833.1990.tb07053.x

Hardy, W.B. (1900) A preliminary investigation of the conditions which determine the stability of irreversible hydrosols. Proceedings of the Royal Society of London 66, 110–125. https://doi.org/10.1098/rspl.1899.0081

Helmholtz, H. (1853) Ueber einige Gesetze der Vertheilung elektrischer Ströme in körperlichen Leiternmit Anwendung auf die thierisch-elektrischen Versuche. Annalen der Physik und Chemie 165, 211–233. https://doi.org/10.1002/andp.18531650603

Henry, D.C. (1931) The cataphoresis of suspended particles. Part I. – The equation of cataphoresis. Proceedings of the Royal Society A 133, 106–129. https://doi.org/10.1098/rspa.1931.0133

Hernández, V.A. (2023) An overview of surface forces and the DLVO theory. ChemTexts 9, 10. https://doi.org/10.1007/s40828-023-00182-9

Hiemenz, P.C., Rajagopalan, R. (1997) Principles of Colloid and Surface Chemistry. 3rd ed., CRC Press, Taylor & Francis Group, Boca Raton.

Hoek, E.M.V., Bhattacharjee, S., Elimelech, M. (2003) Effect of membrane surface roughness on colloid-membrane DLVO interactions. Langmuir 19, 4836–4847. https://doi.org/10.1021/la027083c

Hsu, J.P., Kuo, Y.-C. (1997) The critical coagulation concentration of counterions: Spherical particles in asymmetric electrolyte solutions. Journal of Colloid and Interface Science 185, 530–537. https://doi.org/10.1006/jcis.1996.4591

Hsu, J.P., Liu, B.T. (1998) Effect of Particle Size on Critical Coagulation Concentration. Journal of Colloid and Interface Science 198, 186–189. https://doi.org/10.1006/jcis.1997.5275

Hunter, R.J. (1981) Zeta Potential in Colloid Science. Principles and Applications. Academic Press.

Hunter, R.J., Leyendekkers, J.V. (1978) Viscoelectric coefficient for water. Journal of the Chemical Society, Faraday Transactions 1 74, 450–455. https://doi.org/10.1039/F19787400450

Hussain, S., Awad, J., Sarkar, B., Chow, C.W.K., Duan, J., van Leeuwen, J. (2019) Coagulation of dissolved organic matter in surface water by novel titanium (III) chloride: Mechanistic surface chemical and spectroscopic characterisation. Separation and Purification Technology 213, 213–223. https://doi.org/10.1016/j.seppur.2018.12.038

Hussain, S., van Leeuwen, J., Chow, C.W.K., Aryal. R., Beecham, S., Duan, J., Drikas, M. (2014) Comparison of the coagulation performance of tetravalent titanium and zirconium salts with alum. Chemical Engineering Journal 254, 635–646. https://doi.org/10.1016/j.cej.2014.06.014

Israelachvili, J.N. (2011) Intermolecular and Surface Forces. Elsevier.

Jackson, P.J., Tomlinson, E.J. (1986) Automatic coagulation control-evaluation of strategies and techniques. Water Supply 4, 55–67.

Jarvis, P., Sharp, E., Pidou, M., Molinder, R., Parsons, S.A., Jefferson, B. (2012) Comparison of coagulation performance and floc properties using a novel zirconium coagulant against traditional ferric and alum coagulants. Water Research 46, 4179–4187. https://doi.org/10.1016/j.watres.2012.04.043

Jiang, J.-Q. (2015) The role of coagulation in water treatment. Current Opinion in Chemical Engineering 8, 36–44. https://doi.org/10.1016/j.coche.2015.01.008

Jiang, X., Li, Y., Tang, X., Jiang, J., He, Q., Xiong, Z., Zheng, H. (2021) Biopolymer-based flocculants: a review of recent technologies. Environmental Science and Pollution Research 28, 46934–46963. https://doi.org/10.1007/s11356-021-15299-y

Johannessen, K. (2012) A nonlinear differential equation related to the Jacobi elliptic functions. International Journal of Differential Equations, Article ID 412569. https://doi.org/10.1155/2012/412569

Kang, L.-S. (1995) Effects of temperature on flocculation kinetics using Fe(lll) coagulant in water treatment. Journal of the Korean Environmental Sciences Society 4, 181–194. https://koreascience.or.kr/article/JAKO199511920672767.page

Keh, H.J., Ding, J.M. (2000) Sedimentation velocity and potential in concentrated suspensions of charged spheres with arbitrary double-layer thickness. Journal of Colloid and Interface Science 227, 540–552. https://doi.org/10.1006/jcis.2000.6918

Kim, K.J., Fane, A.G., Nystrom, M., Pihlajamaki, A., Bowen, W.R., Mukhtar, H. (1996) Evaluation of electroosmosis and streaming potential for measurement of electric charges of polymeric membranes. Journal of Membrane Science 116, 149–159. https://doi.org/10.1016/0376-7388(96)00038-5

Kurniawan, S.B., Imron, M.F., Chik, C.E.N.C.E., Owodunni, A.A., Ahmad, A., Alnawajha, M.M., Rahim, N.F.M., Said, N.S.M., Abdullah, S.R.S., Kasan, N.A., Ismail, S., Othman, A.R., Hasan, H.A. (2022) What compound inside biocoagulants/bioflocculants is contributing the most to the coagulation and flocculation processes? Science of the Total Environment 806, 150902. https://doi.org/10.1016/j.scitotenv.2021.150902

La Mer, V.K., Healy, T.W. (1963) The role of filtration in investigating flocculation and redispersion of colloidal dispersions. Journal of Physical Chemistry 67, 2417–2420. https://doi.org/10.1021/j100805a038

Langelier, W.F. (1921) Coagulation of water with alum by prolonged agitation. Engineering News-Record 86, 924–928.

Langelier, W.F., Ludwig, H.F. (1949) Mechanism of flocculation in the clarification of turbid waters. Journal of the American Water Works Association 41, 163–181. https://doi.org/10.1002/j.1551-8833.1949.tb18692.x

Li, L., Rong, S., Wang, R., Yu, S. (2021) Recent advances in artificial intelligence and machine learning for nonlinear relationship analysis and process control in drinking water treatment: A review. Chemical Engineering Journal 405, 126673. https://doi.org/10.1016/j.cej.2020.126673

Liu, J., Long, Y., Zhu, G., Hursthouse, A.S. (2024) Application of artificial intelligence in the management of coagulation treatment engineering system. Processes 12, 1824. https://doi.org/10.3390/pr12091824

Lyklema, J. (2010) Molecular interpretation of electrokinetic potentials. Current Opinion in Colloid & Interface Science 15, 125–130. http://dx.doi.org/10.1016/j.cocis.2010.01.001

Lyklema, J. (2013) Coagulation by multivalent counterions and the Schulze-Hardy rule. Journal of Colloid and Interface Science Volume 392, 102–104. https://doi.org/10.1016/j.jcis.2012.09.066

Lyklema, J. Overbeek, J.T.G. (1961) On the interpretation of electronic potentials. Journal of Colloid Science 16, 501–512. https://doi.org/10.1016/0095-8522(61)90029-0

Maćczak, P., Kaczmarek, H., Ziegler-Borowska, M. (2020) Recent achievements in polymer bio-based flocculants for water treatment. Materials 2020, 3951. https://doi.org/10.3390/ma13183951

Malmberg, C.G., Maryott, A.A. (1956) Dielectric constant of water from 0 °C to 100 °C. Journal of Research of the National Bureau of Standards 56, 2641. https://nvlpubs.nist.gov/nistpubs/jres/56/jresv56n1p1_a1b.pdf

Marlow, B.J., Rowell, R.L. (1985) Sedimentation potential in aqueous electrolytes. Langmuir 1, 83–90. https://doi.org/10.1021/la00061a013

Masliyah, J., Bhattacharjee, S. (2006) Electrokinetic and Colloid Transport Phenomena. Wiley.

Mattson, S. (1928) Cataphoresis and the electrical neutralization of colloidal material. Journal of Physical Chemistry 32, 1532–1552. https://doi.org/10.1021/j150292a011

Mohtadi, M.F., Rao, P.N. (1973) Effect of temperature on flocculation of aqueous dispersions. Water Research 7, 747–767. https://doi.org/10.1016/0043-1354(73)90091-2

Morris, J.K., Knocke, W.R. (1984) Temperature effects on the use of metal-ion coagulants for water treatment. Journal of the American Water Works Association 76, 74–79. https://www.jstor.org/stable/41271959

Ohshima, H. (1994) A simple expression for Henry's function for the retardation effect in electrophoresis of spherical colloidal particles. Journal of Colloid and Interface Science 168, 269–271. https://doi.org/10.1006/jcis.1994.1419

Ohshima, H. (2019) Electrokinetic phenomena in a dilute suspension of spherical solid colloidal particles with a hydrodynamically slipping surface in an aqueous electrolyte solution. Advances in Colloid and Interface Science 272, 101996. https://doi.org/10.1016/j.cis.2019.101996

Ohshima, H., Healy, T.W., White, L.R., O'Brien, R.W. (1984) Sedimentation velocity and potential in a dilute suspension of charged spherical colloidal particles. Journal of the Chemical Society, Faraday Transactions 2: Molecular and Chemical Physics 80, 1299–1317. https://doi.org/10.1039/F29848001299

Okoro, B.U., Sharifi, S., Jesson, M.A., Bridgeman, J. (2021) Natural organic matter (NOM) and turbidity removal by plant-based coagulants: A review. Journal of Environmental Chemical Engineering 9, 106588. https://doi.org/10.1016/j.jece.2021.106588

Plohl, O., Zemljič, L.F., Potrč, S., Luxbacher, T. (2020) Applicability of electro-osmotic flow for the analysis of the surface zeta potential. Royal Society of Chemistry Advances 10, 6777–6789. https://doi.org/10.1039/C9RA10414C

Polat, M., Polat, H. (2010) Analytical solution of Poisson–Boltzmann equation for interacting plates of arbitrary potentials and same sign. Journal of Colloid and Interface Science 341, 178–185. https://doi.org/10.1016/j.jcis.2009.09.008

Ratnaweera, H., Fettig, J. (2015) State of the art of online monitoring and control of the coagulation process. Water 7, 6574–6597. https://doi.org/10.3390/w7116574

Reuss, F. F. (1809) Mémoires de la Societé Impériale des Naturalistes de Moscou 2, 327–337. https://www.biodiversitylibrary.org/item/234594#page/375/mode/1up

Sakong, S., Gross, A. (2018) The electric double layer at metal–water interfaces revisited based on a charge polarization scheme. Journal of Chemical Physics 149, 084705. https://doi.org/10.1063/1.5040056

Saleem, M., Bachmann, R.T. (2019) A contemporary review on plant-based coagulants for applications in water treatment. Journal of Industrial and Engineering Chemistry 72, 281–297. https://doi.org/10.1016/j.jiec.2018.12.029

Šamaj, L., Trizac, E. (2019) Electric double layers with surface charge modulations: Exact Poisson-Boltzmann solutions. Physical Review E 100, 042611. https://doi.org/10.1103/PhysRevE.100.042611

Schulze, H. (1882) Schwefelarsen in wässriger Lösung. Journal für Praktische Chemie 25, 431–452. https://doi.org/10.1002/prac.18820250142

Smoluchowski, M. (1903) Contribution à la théorie de l'endosmose électrique et de quelques phénomènes corrélatifs. Bulletin de l'Académie des Sciences de Cracovie 182–199. https://publikationen.ub.uni-frankfurt.de/frontdoor/index/index/docId/20308

Smoluchowski, M. (1921) In: Handbuch der Electrizität und des Magnetismus, ed. Graetz, L., VEB Georg Thieme: Barth, Leipzig, Vol II, p. 385.

Stern, O. (1924) Zur Theorie der elektrolytischen Doppelschicht. Zeitschrift für Elektrochemie 30, 508–516. https://doi.org/10.1002/bbpc.192400182

Szymczyk, A., Fievet, P., Mullet, M., Reggiani, J.C., Pagetti, J. (1998) Comparison of two electrokinetic methods – electroosmosis and streaming potential – to determine the zeta-potential of plane ceramic membranes. Journal of Membrane Science 143, 189–195. https://doi.org/10.1016/S0376-7388(97)00340-2

Tassinari, B., Conaghan, S., Freeland, B., Marison, W. (2015) Application of turbidity meters for the quantitative analysis of flocculation in a jar test apparatus. Journal of Environmental Engineering 141, 04015015. https://doi.org/10.1061/(ASCE)EE.1943-7870.0000940

TeKippe, R.J., Ham, R.K. (1970a) Coagulation testing: A comparison of techniques – Part I. Journal of the American Water Works Association 62, 594–602. https://doi.org/10.1002/j.1551-8833.1970.tb03972.x

TeKippe, R.J., Ham, R.K. (1970b) Coagulation testing: A comparison of techniques – Part II. Journal of the American Water Works Association 62, 620–628. https://doi.org/10.1002/j.1551-8833.1970.tb03980.x

Tesch, R., Kowalski, P.M., Eikerling, M.H. (2021) Properties of the Pt(111)/electrolyte electrochemical interface studied with a hybrid DFT–solvation approach. Journal of Physics: Condensed Matter 33, 444004. http://dx.doi.org/10.1088/1361-648X/ac1aa2

Travesset, A. (2024) Nonlinear Poisson–Boltzmann solutions for charged parallel plates: When opposite charges repel. Journal of Chemical Physics 161, 054903. https://doi.org/10.1063/5.0221826

Trefalt, G. (2016) Derivation of the inverse Schulze-Hardy rule. Physical Review E 93, 032612. https://doi.org/10.1103/PhysRevE.93.032612

Trefalt, G., Ruiz-Cabello, F.J.M., Borkovec, M. (2015) Interaction forces, heteroaggregation, and deposition involving charged colloidal particles. Journal of Physical Chemistry B 118, 6346–6355. https://doi.org/10.1021/jp503564p

Trefalt, G., Szilagyi, I., Borkovec, M. (2013) Poisson-Boltzmann description of interaction forces and aggregation rates involving charged colloidal particles in asymmetric electrolytes. Journal of Colloid and Interface Science 406, 111–120. https://doi.org/10.1016/j.jcis.2013.05.071

Trefalt, G., Szilágyi, I., Borkovec, M. (2020) Schulze-Hardy rule revisited. Colloid and Polymer Science 298, 961–967. https://doi.org/10.1007/s00396-020-04665-w

Trefalt, G., Szilágyi, I., Téllez, G., Borkovec, M. (2017) Colloidal stability in asymmetric electrolytes: modifications of the Schulze–Hardy rule. Langmuir 33, 1695–1704. https://doi.org/10.1021/acs.langmuir.6b04464

Tzoupanos, N.D., Zouboulis, A.I. (2008) Coagulation-flocculation processes in water/wastewater treatment: The application of new generation of chemical reagents. Proc. 6th IASME/WSEAS International Conference on Heat Transfer, Thermal Engineering and Environment, Rhodes, Greece, August 20–22, 2008, pp. 309–317. https://api.semanticscholar.org/CorpusID:13896581

Tzoupanos, N.D., Zouboulis, A.I., Tsoleridis, C.A. (2009) A systematic study for the characterization of a novel coagulant (polyaluminium silicate chloride). Colloids and Surfaces A: Physicochemical and Engineering Aspects 342, 30–39. https://doi.org/10.1016/j.colsurfa.2009.03.054

Wiersema, P.H., Loeb, A.L., Overbeek, J.T.G. (1966) Calculation of the electrophoretic mobility of a spherical colloid particle. Journal of Colloid and Interface Science 22, 78–99. https://doi.org/10.1016/0021-9797(66)90069-5

Wu, G.-D., Lo, S.-L. (2008) Predicting real-time coagulant dosage in water treatment by artificial neural networks and adaptive network-based fuzzy inference system. Engineering Applications of Artificial Intelligence 21, 1189–1195. https://doi.org/10.1016/j.engappai.2008.03.015

Xiao, F., Huang, J.-C.H., Zhang, B.-J., Cui, C.-W. (2009) Effects of low temperature on coagulation kinetics and floc surface morphology using alum. Desalination 237, 201–213. http://dx.doi.org/10.1016/j.desal.2007.12.033

Xiao, F., Lam, K.M., Li, X.Y., Zhong, R.S., Zhang, X.H. (2011) PIV characterisation of flocculation dynamics and floc structure in water treatment. Colloids and Surfaces A Physicochemical and Engineering Aspects 379, 27–35. http://doi.org/10.1016/j.colsurfa.2010.11.053

Xiao, F., Ma, J., Yi, P., Huang, J.-C.H. (2008) Effects of low temperature on coagulation of kaolinite suspensions. Water Research 42, 2983–2992. https://doi.org/10.1016/j.watres.2008.04.013

Xing, X.J. (2011) Poisson-Boltzmann theory for two parallel uniformly charged plates. Physical Review E 83, 041410. https://doi.org/10.1103/PhysRevE.83.041410

Yamaguchi, N.U., Cusioli, L.F., Quesada, H.B., Ferreira, M.E.C., Fagundes-Klen, M.R., Vieira, A.M.S., Gomes, R.G., Vieira, M.F., Bergamasco, R. (2021) A review of Moringa oleifera seeds in water treatment: Trends and future challenges. Process Safety and Environmental Protection 147, 405–420. https://doi.org/10.1016/j.psep.2020.09.044

Yang, R., Li, H., Huang, M., Yang, H., Li, A. (2016) A review on chitosan-based flocculants and their applications in water treatment. Water Research 95, 59–89. http://dx.doi.org/10.1016/j.watres.2016.02.068

Yang, R.-J., Fu, L.-M., Hwang, C.-C. (2001) Electroosmotic entry flow in a microchannel. Journal of Colloid and Interface Science 244, 173–179. https://doi.org/10.1006/jcis.2001.7847

Yu, W.-Z., Liu, T., Gregory, J., Li, G.-B., Liu, H.-J., Qu, J.-H. (2012) Aggregation of nano-sized alum–humic primary particles. Separation and Purification Technology 99, 44–49. http://dx.doi.org/10.1016/j.seppur.2012.08.017

Zeng, D., Wu, J., Kennedy, J.F. (2008) Application of a chitosan flocculant to water treatment. Carbohydrate Polymers 71, 135–139. doi:10.1016/j.carbpol.2007.07.039

Zhang, B., Tang, H., Huang, D., Gao, X., Zhang, B., Shen, Y., Shi, W. (2020) Effect of pH on anionic polyacrylamide adhesion: New insights into membrane fouling based on XDLVO analysis. Journal of Molecular Liquids 320, 114463. https://doi.org/10.1016/j.molliq.2020.114463

Zhang, W., Wang, Q., Zeng, M., Zhao, C. (2018) An exact solution of the nonlinear Poisson-Boltzmann equation in parallel-plate geometry. Colloid and Polymer Science 296, 1917–1923. https://doi.org/10.1007/s00396-018-4394-8

Zhao, C., Yang, C. (2013) Electrokinetics of non-Newtonian fluids: A review. Advances in Colloid and Interface Science 201–202, 94–108. http://dx.doi.org/10.1016/j.cis.2013.09.001

Zhao, Y.X., Gao, B.Y., Shon, H.K., Cao, B.C., Kim, J.-H. (2011) Coagulation characteristics of titanium (Ti) salt coagulant compared with aluminum (Al) and iron (Fe) salts. Journal of Hazardous Materials 185, 1536–1542. https://doi.org/10.1016/j.jhazmat.2010.10.084

Zhuang, J., Qi, Y., Yang, H., Li, H., Shi, T. (2021) Preparation of polyaluminum zirconium silicate coagulant and its performance in water treatment. Journal of Water Process Engineering 41, 102023. https://doi.org/10.1016/j.jwpe.2021.102023

Chapter 3
Process of drinking water treatment using coagulation/flocculation method

The process of drinking water treatment is non-uniform not only what concerns the locations of the water treatment plants (WTPs) but also significantly subjects to the individual seasons round the year. In addition, other factors contributing to a variety of input conditions in the WTPs are caused by (un)predictable meteorological abrupt changes such as flash floods or landscape changes (building of new municipal districts or industrial complexes, cultivation of agricultural land, and livestock production).

All these changes are interlaced with new challenges the WTPs have to face. It means that alongside with an existence of classical impurities in water as natural organic matters (NOMs) and suspended particles (clay, silt, sand, etc.), the WTPs are also confronted with the presence of, for instance, pesticides, pharmaceuticals, and plastic nano- and micro-particles. Tourism and fish farming also participate in the adverse impacts on the quality of surface water used in the WTPs.

However, the production of drinking water is influenced not only by all these unfavourable inputs present at the initial stage of water treatment but also by the unwanted products emerging during the treatment process itself: the so-called disinfection by-products (DBP) possibly posing a health risk arising during the stage of chlorination, chloramination, or ozonation (chlorine and chloramine react with decaying organic matter).

At the very beginning of the coagulation/flocculation process there is a mutual interplay between three factors: a proper choice of a coagulant agent and additives (flocculants), its (their) dosage, and equally important, an adequate choice of pH number. An adequate choice of the pH number even enables – to some extent – to balance temperature changes; otherwise lower efficiency with a decreasing temperature (Hanson and Cleasby 1990; Van Benschoten and Edzwald 1990; Kang and Cleasby 1995).

A determination of the optimal pH value is not an easy procedure not only from the physico-chemical viewpoint but also from its introduction as discussed below.

3.1 Density of measurements – pH vs. concentration of ions H$^+$

A potential of hydrogen (pH) number is not a physico-chemical quantity, the pH number is just the numerical transformation of equilibrium molar concentration of ions H$^+$ [H$^+$]

$$pH = -\log_{10}([H^+]),\qquad(3.1)$$

https://doi.org/10.1515/9783111246765-003

as introduced by Sørensen (1909). The problem is that the nonlinear transformation uses a common logarithmic function. This means that the logarithmic images of two numbers can be taken nearly as two 'neighbours'; however the original values (numbers) (in this case a difference in [H$^+$]) can be really remoted. To document this situation, we can depict a selected set of equidistant numbers {6, 6.5, 7, 7.5, 8} in the interval [6,8] which is amply used in testing a suitable pH number for an application in water treatment (see Fig. 3.1 (left)). If the data are transformed back to their original meaning (concentration [H$^+$]) (see Fig. 3.1 (right)), we can see that a region for pH between 6 and 7.5 covering three fourth of the whole area between pH = 6 and 8 is now shrunk to lower than one third if depicted in real concentration values between 31.6×10^{-8} and 100×10^{-8} ($\equiv 10^{-6}$). The square boxes in Fig. 3.1 represent e.g. the percentage values of turbidity or dissolved organic carbon (DOC) as two principal indicators of water quality. In practice, it means that a surface expressing a course of turbidity or DOC is generated by 24 experimental points over less than one third of the whole area and by only 12 (boundary) points over more than two third of the whole area. This implies that there is a high probability of losing possible extreme characteristics in the discussed region. It is illustrated in Fig. 3.2, where a relatively flat plain (top) has a lot of space between the neighbouring experimental points (pH = 6 and 6.5) to convert to a concave (middle) or convex (bottom) profile. There is a possibility to attenuate such excess by balancing the experimental points with respect to concentration [H$^+$]. Partitioning of rounded pH values as for instance indicated in Fig. 3.3 ensures practically equidistant differences in equilibrium molar concentration of ions H$^+$ [H$^+$]. If we prolong the studied interval [6,8] to wider intervals [5,8] or [6,9] the situation gets even worse. Sudden changes then require denser local partitioning. In the case of denser partitioning there can be a question how precise the value of pH can be prepared in practice.

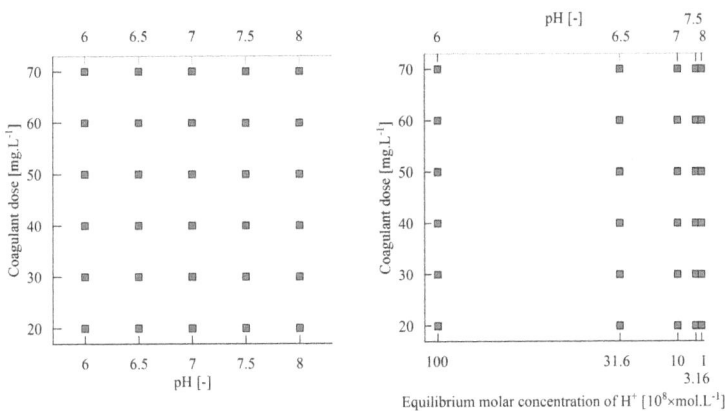

Fig. 3.1: Imbalance in uniformity of the experimental grids using logarithmic (left) and normal (right) scales for equilibrium molar concentration of H$^+$.

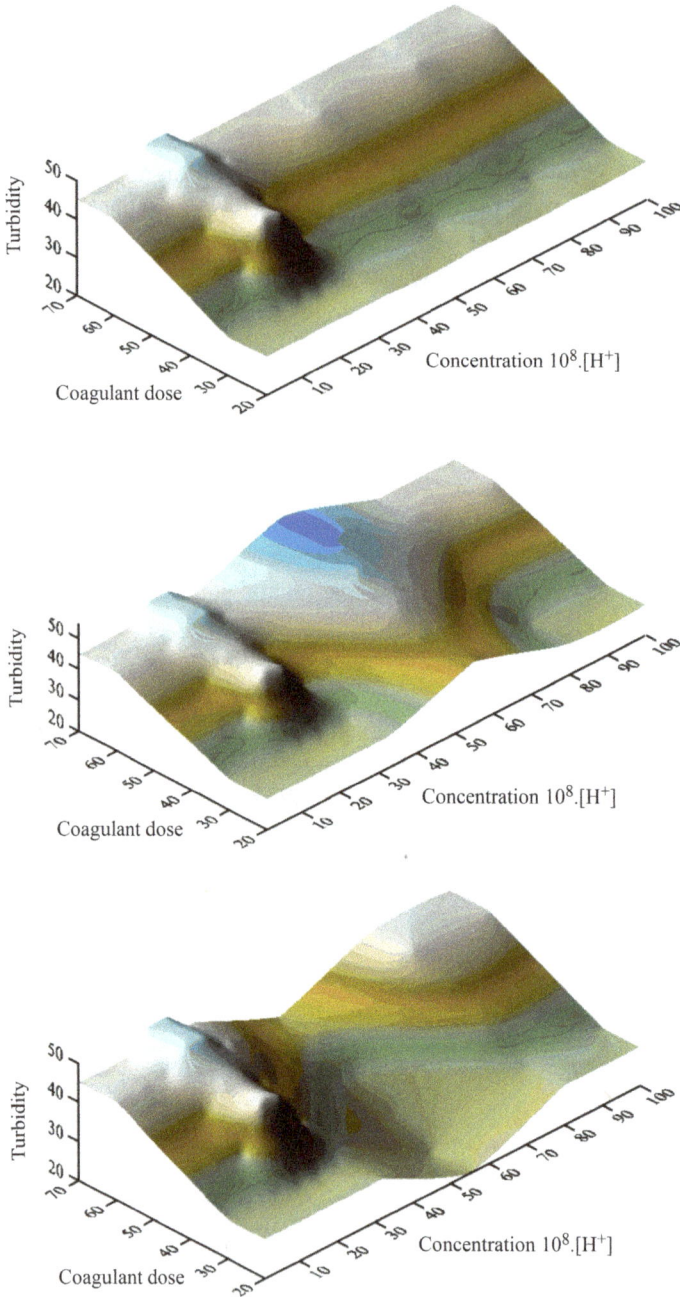

Fig. 3.2: Possible surface profiles – flat (top), concave (middle), convex (bottom) – between neighbouring experimental points at pH = 6.5 and 6 (corresponding to 31.6×10^{-8} and 100×10^{-8} if expressed in [H$^+$]).

Equilibrium molar concentration of H^+ [$10^8 \times$mol.L^{-1}]

100	79.4	50.1	25.1	1

6	6.1	6.3	6.6	8

pH [-]

6	6.05	6.1	6.15	6.2	6.3	6.4	6.5	6.7	7	8

100	89.1	79.4	70.8	63.1	50.1	39.8	31.6	20	10	1

Equilibrium molar concentration of H^+ [$10^8 \times$mol.L^{-1}]

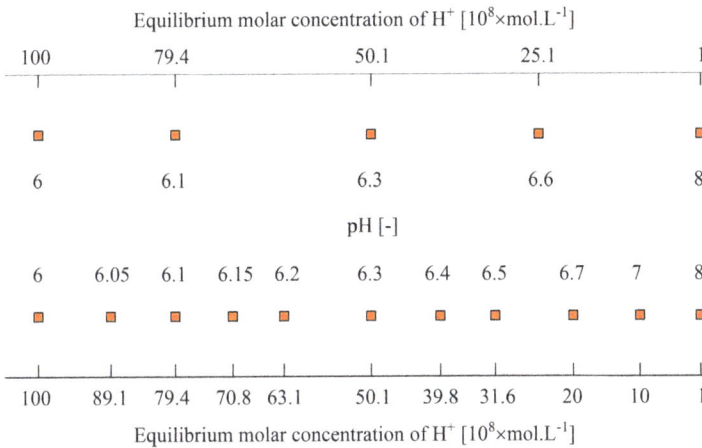

Fig. 3.3: Nearly equidistant (sparse and dense) partitioning of the studied logarithmic interval [6,8] with respect to real values ([H^+]).

3.2 Natural organic matters (NOMs)

The natural organic matter (NOM) as a ubiquitous component both in the lithosphere and hydrosphere (Kosobucki and Buszewski 2014) represents a huge reservoir of the carbon in the environment. In principle, NOM is non-toxic; however, it strongly contributes to the significant deterioration of organoleptic properties of water (colour, taste, and odour), to a reduction in the amount of dissolved oxygen in water, and its presence enhances forming of harmful DBP. Due to a composition of NOM, an amount of dissolved organic carbon (DOC) serves as an optimal indicator of NOM in drinking water.

NOM is created by decaying flora and fauna and hence is very complex heterogeneous organic material. It can be divided into two basic groups: dissolved and particulate organic matter (Abu Hasan et al. 2020), the first one passing through a sieve with the mesh 0.45 × 0.45 µm, the second one is retained. Dissolved organic matter is either hydrophobic (humic substances traditionally categorised according to their solubility – humic acid, fulvic acid and humin) or hydrophilic (e.g. aliphatic carbon and nitrogenous compounds – carboxylic acids, carbohydrates, and proteins). Out of these components, humic substances participate in more than one half of the total organic carbon content in water.

There is a series of the methods how to remove NOM from the treated water: coagulation/flocculation, adsorption, advanced oxidation methods, membrane filtration, biological degradation, ion exchange, and others (Zhang et al. 2015; Ciobanu et al. 2022). Their choice subjects to their efficiency (for a specific raw water), financial expenses, availability, required water amount, an impact on the environment, etc. The most frequent (and economically feasible) method is traditionally coagulation/flocculation with consequent sedimentation/flotation and sand filtration (Jacangelo et al. 1995, Matilainen et al. 2010).

From the viewpoint of water treatment a very important issue is a proper evaluation of the NOM characteristics (as e.g. DOC and nitrogen, (specific) ultraviolet absorbance, fluorescence, polarity, zeta potential, chemical oxygen demand, pH, turbidity, and colour). The summary of available experimental devices determining these characteristics is presented in Matilainen et al. (2011), Pan et al. (2016), and Chen et al. (2021).

In the recent period, the traditional seasonal variations in NOM (Sharp et al. 2006) are also accompanied by the climate changes resulting in increasing concentration and changing character of NOM towards predominantly hydrophobic nature (Anderson et al. 2023).

Not so intensive in the centre of attention is an impact of residual NOM on the distribution system concerning among other things the growth of biofilms (Jacangelo et al. 1995; Riyadh et al. 2024).

3.3 Hydrolysis of metal cations

According to the Schulze-Hardy rule (Section 2.2) higher valency of the metal cations (e.g. trivalent Al^{3+} and Fe^{3+}) should contribute to better destabilisation of negatively charged impurities in water. The situation is not so unambiguous due to immediate hydrolysis of the metal cations in water which dominantly changes their behaviour from the viewpoint of coagulant agents.

In water surroundings, the cations of Al^{3+} and Fe^{3+} are encircled by six water molecules resulting from polar nature of water. Relatively high positive charge of a centrally located metal cation induces an orientation of six water molecules in the primary hydration shell with the electrons closer to the cation (see Fig. 3.4 (left)). It can result in the dissociation of a proton H^+, which leaves a hydroxyl group and thus contributes to the reduction of positive charge of the central cation (see Fig. 3.4 (right)). Due to splitting of the water molecules the process is denoted as hydrolysis. It is apparent that hydrolysis subjects to pH values of media, and dissociation is enhanced with higher pH values and vice versa.

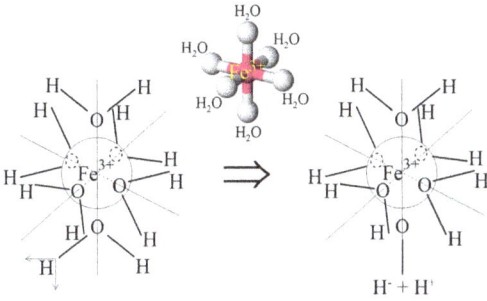

Fig. 3.4: Six water molecules in octahedral coordination round Fe^{3+} cation (left), dissociation of a proton H^+ (right).

The following notation (simplified by not introducing the water molecules in the primary hydration shell)

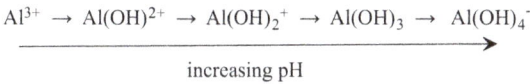

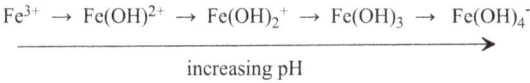

$$Al^{3+} \rightarrow Al(OH)^{2+} \rightarrow Al(OH)_2^+ \rightarrow Al(OH)_3 \rightarrow Al(OH)_4^-$$

increasing pH

$$Fe^{3+} \rightarrow Fe(OH)^{2+} \rightarrow Fe(OH)_2^+ \rightarrow Fe(OH)_3 \rightarrow Fe(OH)_4^-$$

increasing pH

describes a successive development of hydrolysis equilibria. In more detail

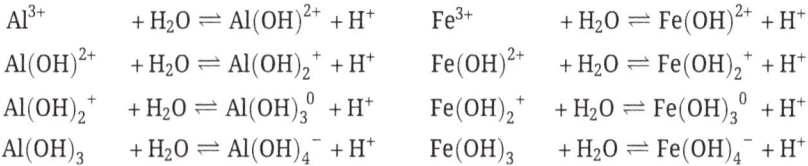

$$Al^{3+} \quad + H_2O \rightleftharpoons Al(OH)^{2+} + H^+ \qquad Fe^{3+} \quad + H_2O \rightleftharpoons Fe(OH)^{2+} + H^+$$
$$Al(OH)^{2+} \quad + H_2O \rightleftharpoons Al(OH)_2^+ + H^+ \qquad Fe(OH)^{2+} \quad + H_2O \rightleftharpoons Fe(OH)_2^+ + H^+$$
$$Al(OH)_2^+ \quad + H_2O \rightleftharpoons Al(OH)_3^0 + H^+ \qquad Fe(OH)_2^+ \quad + H_2O \rightleftharpoons Fe(OH)_3^0 + H^+$$
$$Al(OH)_3 \quad + H_2O \rightleftharpoons Al(OH)_4^- + H^+ \qquad Fe(OH)_3 \quad + H_2O \rightleftharpoons Fe(OH)_4^- + H^+$$

where each stage is characterised by its equilibrium constant given by

$$K_3^{Al} = \frac{[Al(OH)_3^0][H^+]}{[Al(OH)_2^+]}.$$

The square brackets represent the equilibrium molar concentrations of the individual species, and the relation follows from the Guldberg and Waage's law of mass action (Waage and Guldberg 1864, Guldberg and Waage 1879) presenting a direct proportionality between the rate of a chemical reaction and the product of the activities or concentrations of the reactants. As the equilibrium constants attain small values, from the practical usage, the introduction of $pK = -\log_{10}K$ using a common logarithm is more comfortable. The determination of the individual equilibrium constants is uneasy and their dispersion introduced in the literature is rather non-negligible. For the

Tab. 3.1: pK_i (i = 1, 2, 3, 4) values for hydrolysis of both metal cations at 25 °C and zero ionic strength (taken from Gregory 2006, Table 6.1).

Metal cation	pK_1	pK_2	pK_3	pK_4
Al^{3+}	4.95	5.6	6.7	5.6
Fe^{3+}	2.2	3.5	6	10

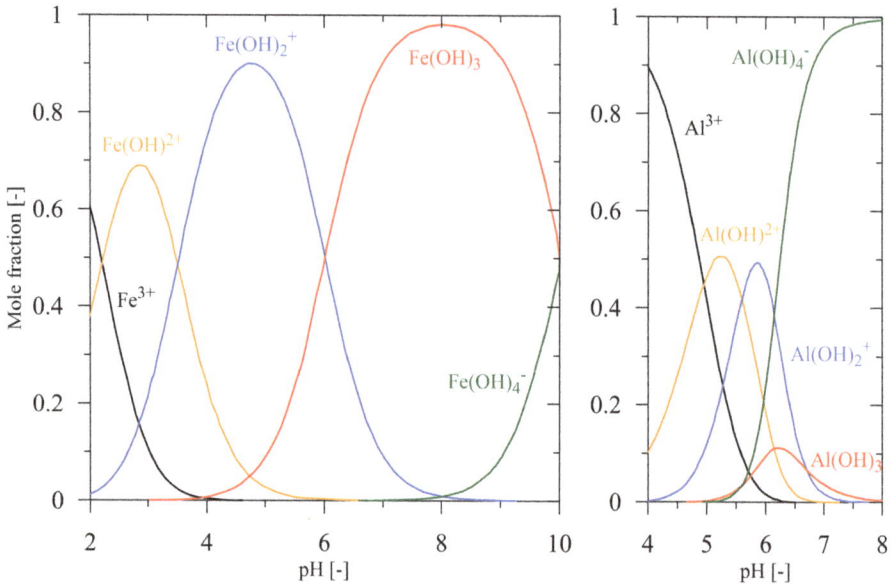

Fig. 3.5: Normalised mole fractions of hydrolysed Fe^{3+} and Al^{3+} species in relation to total soluble metal concentration (summation of all five fractions equals 1).

values given in Gregory (2006) (see Tab. 3.1), the mutual ratios of the hydrolysed species for Fe^{3+} and Al^{3+} are depicted in Fig. 3.5. It documents an apparent difference in mutual participation of the Fe^{3+} individual species in the whole hydrolysed spectrum on the one side and of the Al^{3+} individual species on the other side.

The uncharged hydroxides $Fe(OH)_3$ and $Al(OH)_3$ exhibit low solubility in water and in the specific ranges of pH values (see Fig. 3.6) are susceptible to create precipitate. These ranges of pH values are important from the viewpoint of water treatment. It is necessary to have in mind that the equilibrium constants are dependent on temperature, which influences a participation of the individual hydrolysed species in Fig. 3.5.

As mentioned above, the determination of equilibrium constants is not unambiguous, and the chosen alternative frequently used values in the literature are presented in Tab. 3.2.

Detailed information about hydrolysis in connection with water treatment is presented in Duan and Gregory (2003), Davis and Edwards (2014), and references therein.

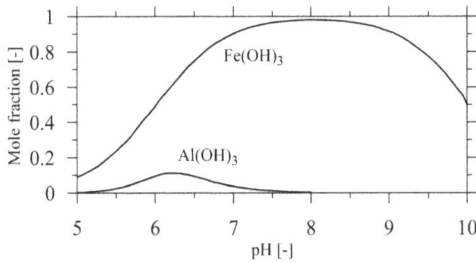

Fig. 3.6: A comparison of mole fractions of hydroxides $Fe(OH)_3$ and $Al(OH)_3$ related to other corresponding Fe and Al-hydrolysed species, respectively.

Tab. 3.2: Equilibrium constants for Al^{3+} and Fe^{3+} cations (data taken from Hummel et al. 2002 (Al^{3+}) and Stefansson 2007 (Fe^{3+})).

Zero ionic strength, temperature 25 °C			
Hydrolysis of Al^{3+} cation	**pK**	**Hydrolysis of Fe^{3+} cation**	**pK**
$Al^{3+} + n\,H_2O \rightleftharpoons Al(OH)_n^{(3-n)} + n\,H^+$		$Fe^{3+} + n\,H_2O \rightleftharpoons Fe(OH)_n^{(3-n)} + n\,H^+$	
$n = 1$	4.96	$n = 1$	2.19
$n = 2$	10.59	$n = 2$	5.76
$n = 3$	16.43	$n = 3$	14.42
$n = 4$	22.88	$n = 4$	21.83
$Al^{3+} + H_2O \rightleftharpoons Al(OH)^{2+} + H^+$	4.96	$Fe^{3+} + H_2O \rightleftharpoons Fe(OH)^{2+} + H^+$	2.19
$Al(OH)^{2+} + H_2O \rightleftharpoons Al(OH)_2^+ + H^+$	5.63	$Fe(OH)^{2+} + H_2O \rightleftharpoons Fe(OH)_2^+ + H^+$	3.57
$Al(OH)_2^+ + H_2O \rightleftharpoons Al(OH)_3^0 + H^+$	5.84	$Fe(OH)_2^+ + H_2O \rightleftharpoons Fe(OH)_3^0 + H^+$	8.66
$Al(OH)_3 + H_2O \rightleftharpoons Al(OH)_4^- + H^+$	6.45	$Fe(OH)_3 + H_2O \rightleftharpoons Fe(OH)_4^- + H^+$	7.41

3.4 An influence of increasing coagulant/flocculant dosage on water characteristics

The course of turbidity curve in dependence on amount of coagulant (alum) dosage is introduced in Tab. 3.3. The natural distribution to four zones corresponds to intensity of particles charge: stability → destabilisation → restabilisation → irreversible destabilisation. A number of zones depends on the value of pH under which the coagulation process is applied; in the case of aluminium sulphate (briefly alum), the full number of zones – four – are achieved for pH values round 7, with an increasing pH value the zone 2 can be reduced or eliminated (for pH = 8) (see Gregory (2006, p. 132)).

Zone 4 is characterised by two simultaneous processes: hydroxide precipitation and sweep coagulation.

Tab. 3.3: A passage from stability to complete destabilisation in dependence on coagulant dosage.

| | Increasing dosage of hydrolysing coagulant | | | |
	→ → → → → → → → → →			
Dosage	Very low	Intermediate	High	Very high
Response	None, stability preserved	Charge neutralisation ⇒ coagulation	Charged reversal, restabilisation	Hydroxide precipitation, sweep coagulation
Zone	Zone 1	Zone 2	Zone 3	Zone 4

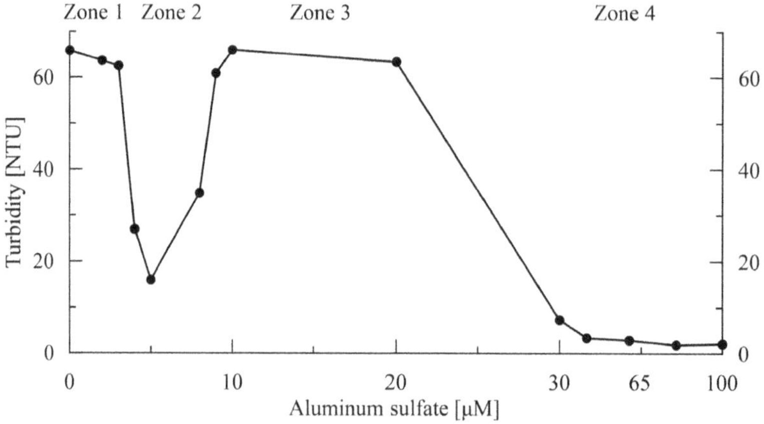

(linearly double-scaled abscissa, data taken from Duan 1997, Fig. 5.4b)

Zone	Zone 1	Zone 2	Zone 3	Zone 4
	Practically no sedimentation, particles still exhibit negative charge	Apparent drop of residual turbidity, charge neutralisation, particles charge around zero, and destabilisation	Process of restabilisation, charge reversal, and flocs breakage	Larger flocs sedimenting more rapidly than in zone 2, no restabilisation occurs

Hydroxide precipitation when soluble metal ions are successively converted to relatively insoluble metal-hydroxide precipitates (such as $Al(OH)_3$ or $Fe(OH)_3$) is discussed above.

The so-called sweep coagulation describes a process when impurity particles are 'gobbled up' by hydroxide precipitate which excessive amount is caused by an application of coagulant in the dosage exceeding solubility of the amorphous hydroxide. The enmeshed particles are removed (swept) from the treated water, even more effectively than using the means of charge neutralisation (e.g. lower dosage). From the other side, this approach results in an apparent increase of sludge formed by precipitating material containing – in an increasing measure – applied metal elements. It means that faster water treatment faces to the problems with sludge processing in the case of aluminium not free of the health risks. Another drawback of Al-based coagulants (in a comparison with the Fe-based ones) is connected with their efficiency at decreasing temperature, and this can be to some extent suppressed by increasing the values of pH (see Davis and Edwards 2014 and references therein). The term sweep coagulation expressing an incorporation of impurities into applied hydrolysed coagulants is logically also denoted as sweep flocculation.

3.5 Basic aggregation mechanisms

The basic aggregation mechanisms reducing (eliminating) NOM from raw water are based on a combination of charge neutralisation, entrapment, adsorption, and complexation (Jarvis et al. 2004, Henderson et al. 2006), some of them already described above.

The process of complexation forming stable complexes is commonly based on non-covalent interactions between two or more compounds (Adusei-Gyamfi et al. 2019).

The process of coagulation/flocculation can be intensified by an application of adsorption procedure (Bhatnagar and Sillanpää 2017). The most frequent adsorbents are granular and powdered activated carbons which inclusion into the treatment process is illustrated in Fig. 3.7. The activated carbons exhibit a large ratio between their surface and volume. Due to porosity on microscopic scale a surface of 1 g exceeds 3,000 m^2. There is also a tendency to replace activated carbons by cheaper and renewable materials (see Menya et al. 2018).

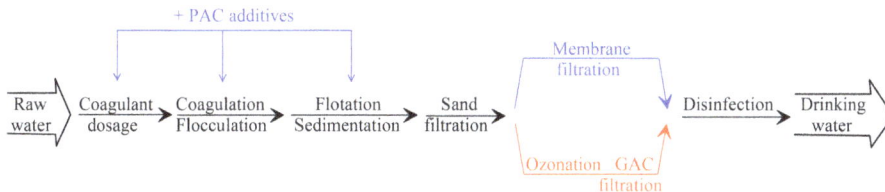

Fig. 3.7: A difference between common application of PAC and GAC in WTPs.

3.6 Traditional frequently used coagulants

The historically most traditional coagulants are based on trivalent aluminium Al^{3+} and iron Fe^{3+} and their position is still relatively fixed also due to the fact that the carboxyl and phenolic groups dominating in NOM have high affinities for metals depending on the pH. The hydrophobic fraction and high molar mass compounds of NOM are removed more efficiently than hydrophilic fraction and the low molar mass compounds (Matilainen et al. 2010).

The classical aluminium sulphate ($Al_2(SO_4)_3 \cdot 18H_2O$, denoted as alum) and ferric sulphate ($Fe_2(SO_4)_3 \cdot 9H_2O$) are often replaced by ferric chloride ($FeCl_3 \cdot 6H_2O$) or pre-polymerised aluminium chloride (abbreviated as PACl or PAC). A jar test (see Fig. 3.8) represents a powerful tool for evaluating efficiency of the agents.

Fig. 3.8: A jar test device for evaluating coagulant efficiency.

Figure 3.9 depicts the efficiency of these four coagulant agents in removing NOM which presence is evaluated by a level of DOC. The experimental data are taken from Pivokonsky et al. (2024) using the same raw water. As apparent, a removal of DOC for Al- and Fe-based coagulants substantially differs. The values of pH corresponding to higher suppression of presence of NOM are shifted for Al-based agents by approximately 1 to higher pH values in comparison with the Fe-based ones, what supports a generally observable phenomenon.

A remarkable disbalance is evident in analysing residuals (see Fig. 3.10). The residuals of Fe more or less exhibit the same low level over the whole wide region of pH (from 4.5 to 10) whereas the residuals of Al attain large values for lower pH, and the situation is not much better for higher values of pH (in this case, the range of pH values was considered between 5 and 8). In this context it is necessary to mention adverse effects of aluminium concerning health risks.

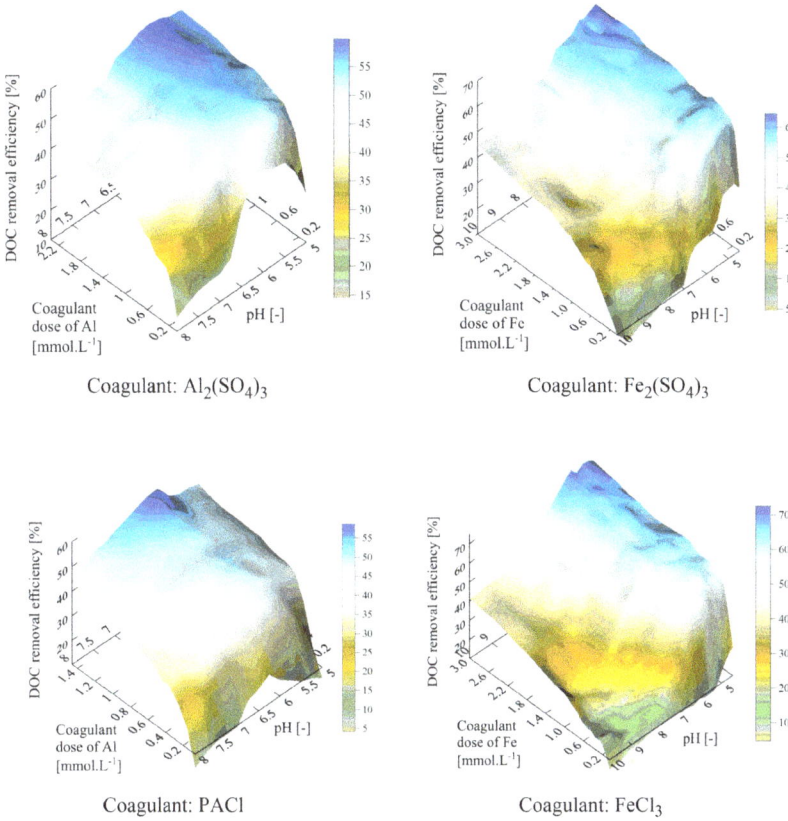

Fig. 3.9: Efficiency of the individual coagulant agents in removing dissolved organic carbon (DOC) (based on the experimental data in Pivokonsky et al. 2024).

Apart from pH values it is also necessary to pay attention to the amounts of coagulants necessary for achieving satisfactory values of DOC removal. To this aim it is useful to chart the contour graphs corresponding to Fig. 3.9. These graphs (see Fig. 3.11) are intentionally charted (as well as the preceding ones) in dependence on the values of recalculated amounts of Al and Fe for all four coagulants. For an immediate comparison, the individual graphs are of uneven heights providing the same position of the individual levels of the applied amounts expressed in molar concentrations. The Fe-based coagulants exhibit higher efficiency for wider range of pH values, the contour for 50% is marked more intensively, better efficiency is achieved at the lighter spots, and darker regions are unacceptable.

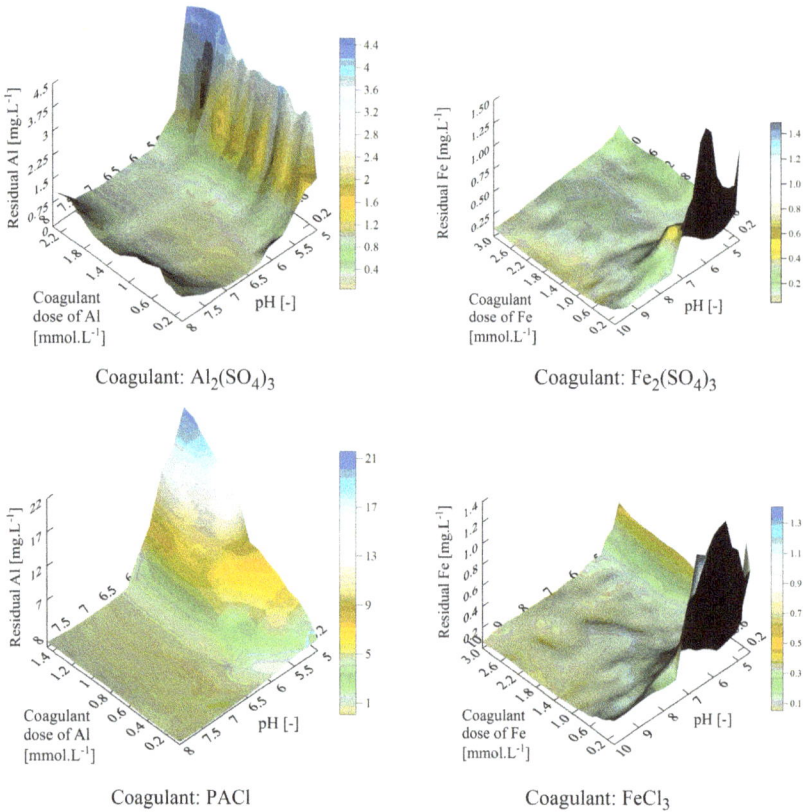

Fig. 3.10: Presence of Al- and Fe-residuals after an application of the individual coagulant agents (based on the experimental data in Pivokonsky et al. 2024).

The contours depicted in Fig. 3.12 express the levels of residual in dependence on pH values and molar concentrations of Al and Fe. The difference between both metals is striking as the regions with levels under 0.2 mmol L^{-1} (a contour with the level 0.2 is thicker) for Fe cover almost all studied interval for pH between 4.5 and 10 contrasting to isolated islands for Al (pH in the range from 5 to 8).

The main conclusion is presented in Fig. 3.13. This figure exemplary explains the principal aim of jar tests. The aim is not to find a local extreme where all parameters fulfil the required criteria but the aim is to find such coagulant agent which exhibits a relatively large plateau fulfilling these criteria. The problem is that in everyday practice in the WTPs the applied coagulant is not (and cannot be) uniformly homogenised. This prefers the agents with relatively flat characterisation of efficiency of the individ-

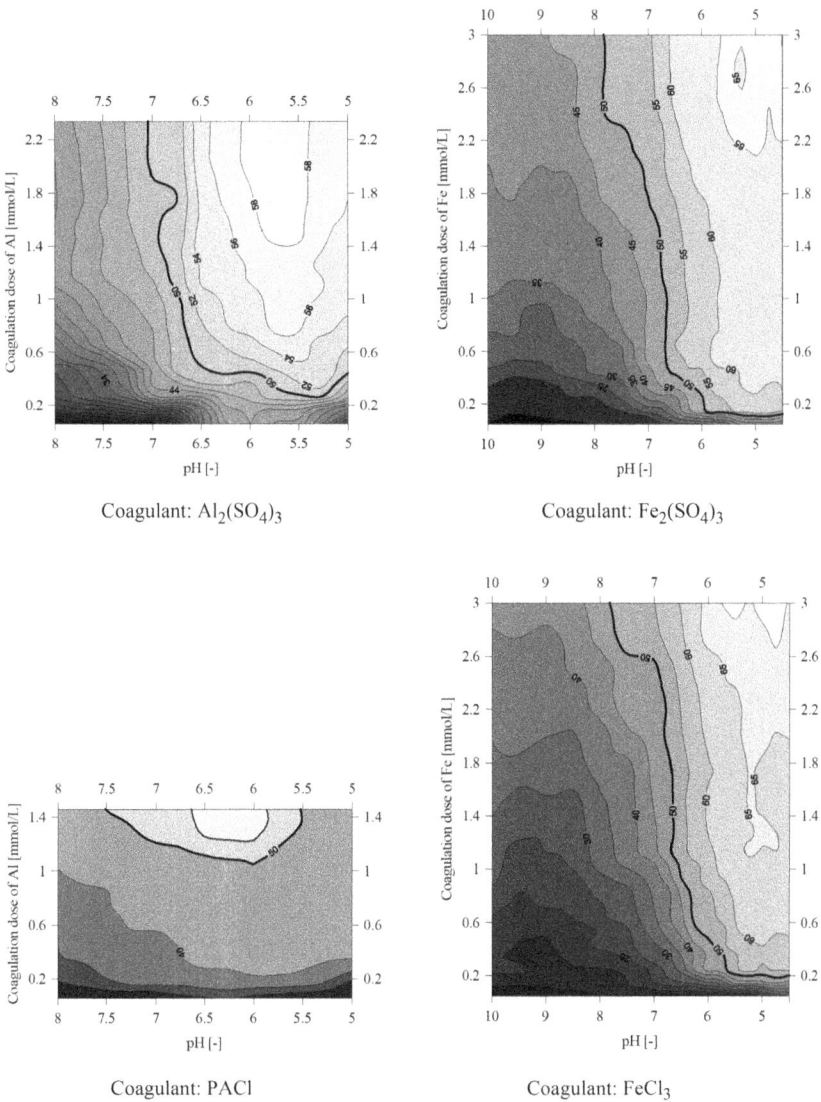

Coagulant: Al$_2$(SO$_4$)$_3$

Coagulant: Fe$_2$(SO$_4$)$_3$

Coagulant: PACl

Coagulant: FeCl$_3$

Fig. 3.11: The contour graphs for the individual coagulant agents expressing a level of DOC removal in % (based on the experimental data in Pivokonsky et al. 2024).

ual items. In the presented experiment an application of PACl seems to be rather inconvenient. It is necessary to point out that these measurements were carried out with one concrete type of raw water, and the results can non-negligibly differ for other types. However, it is possible to state (based on various literature sources) that Fe-based coagulants are more efficient than the Al-based ones (see e.g. Matilainen

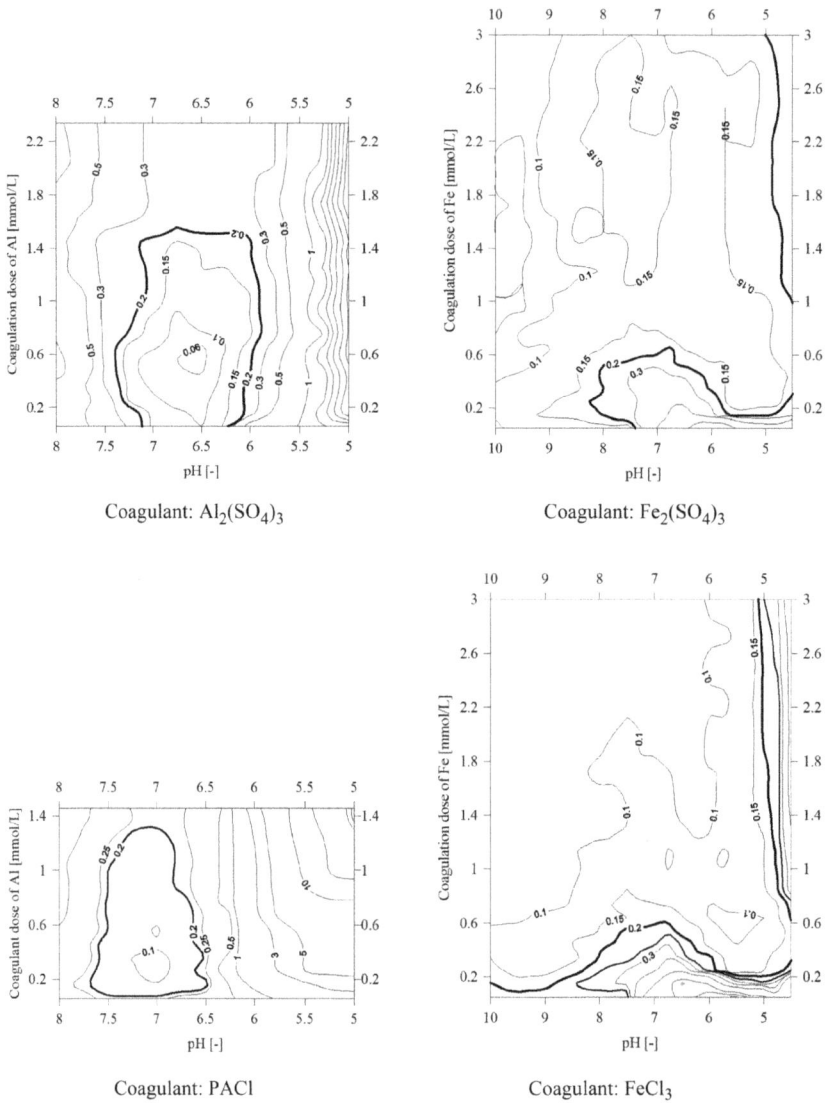

Coagulant: Al$_2$(SO$_4$)$_3$

Coagulant: Fe$_2$(SO$_4$)$_3$

Coagulant: PACl

Coagulant: FeCl$_3$

Fig. 3.12: The contour graphs for the individual coagulant agents expressing a level of residual in mg L^{-1} (based on the experimental data in Pivokonsky et al. 2024).

et al. 2005). Jarvis et al. (2008) show that the ferric coagulants produce stronger flocs than the aluminium ones.

More information on coagulation/flocculation process including a list of used coagulants can be found in the reviews by Edzwald (1993), Matilainen et al. (2010) covering the time period 2006–2009, and Sillanpää et al. (2018) covering the time period

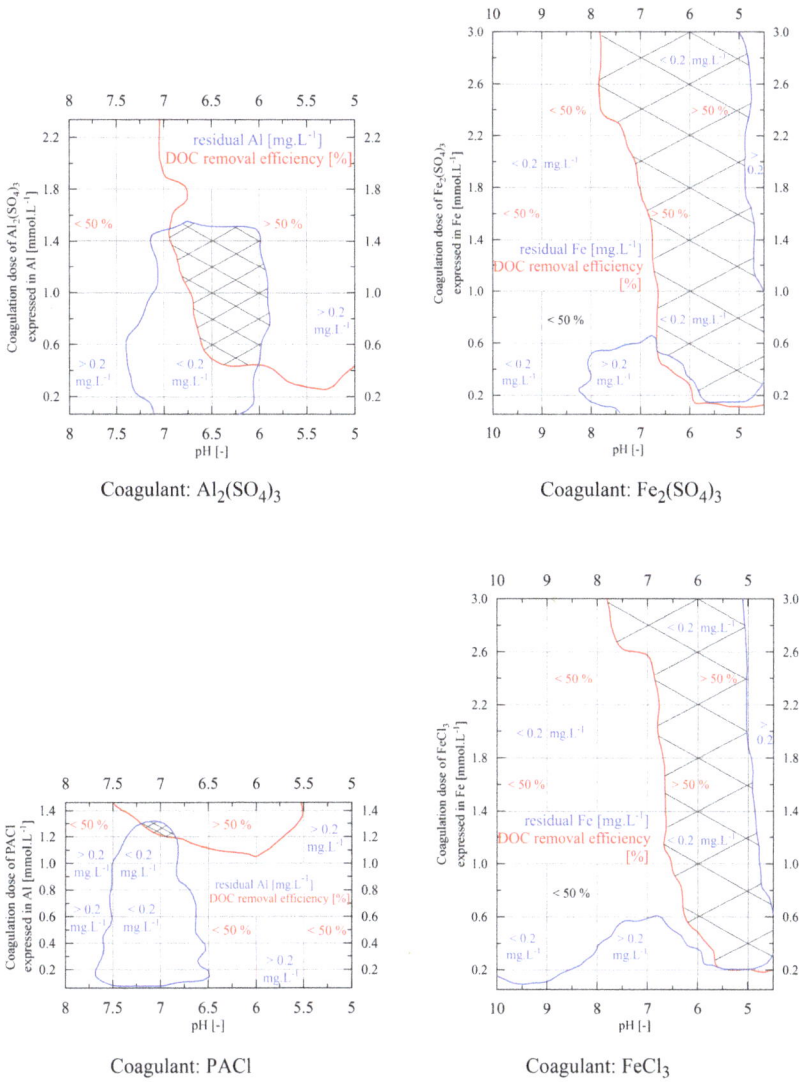

Coagulant: Al$_2$(SO$_4$)$_3$

Coagulant: Fe$_2$(SO$_4$)$_3$

Coagulant: PACl

Coagulant: FeCl$_3$

Fig. 3.13: The optimal regions (shaded) from the viewpoint of DOC removal (>50%) and amount of residual (<0.2 mg L^{-1}) (based on the experimental data in Pivokonsky et al. 2024).

2010–2016. Very efficient is also an application of high molecular weight polymers. Their behaviour is described in Chapter 1.

3.7 Response surface method

A jar-test method represents a traditional approach to evaluating the basic inputs into a coagulation/flocculation process: an amount of coagulant dose, setting of corresponding pH value, and an amount of coagulant/flocculant dosage. In spite of the fact that this method is not standardised (see Section 2.4.1) it provides relatively good approximation of the inputs in question. However, the procedure is time-consuming.

Traditionally more or less equidistant experimental points are chosen in each studied input, and the whole pattern about efficiency is based on consecutive testing of one variable input while the others are fixed. After optimising one input, an optimisation procedure continues with the following one. This trial-and-error approach, also known as the one factor at a time (OFAT) method, pays an emphasis to the individual inputs and rather ignores the interactions between the optimised inputs. It may result in yielding false optimum conditions. Moreover, a number of experiments are not negligible.

One way how to accelerate an optimisation of the input values is an application of the so-called response surface method (RSM, Montgomery 2013, Chapter 11; Khuri and Cornell 2018). This method in a combination with a central composite design (CCD) exhibits two advantages over the classical jar testing:
1) significantly reduces a number of combinations of the input values (operating variables) necessary for the determination of the optimised values;
2) takes into account the interactions among the studied quantities.

The distribution of operating variables using the CCD method depends on a number k of entry factors (such as coagulant dose, pH value, and flocculant dose) (see Fig. 3.14). The individual experimental points (combination of input variables) are usually located symmetrically with respect the central point ([0,0] in the case of a circle (its perimeter) or [0,0,0] in the case of a sphere (its surface). The values $-a$, -1, 0, $+1$, and $+a$ represent the so-called coded values corresponding to the actual values of the individual input variables (see Fig. 3.15 and Tab. 3.4). The range of actual values for a specific factor X_i spans from X_{min} (corresponding to $-a$) to X_{max} (corresponding to $+a$) and the intermediate factor values for -1, 0, $+1$ are calculated in a linear way. As the limiting values for each factor (X_{min}, X_{max}) are attributed to the coded values with a at different positions, a number of coded values containing a in the CCD always double a number of factors.

A value of a can be chosen as $a = 1$ (and sometimes is taken at this value) in the case when two factors are considered (see Fig. 3.16 and Tabs. 3.5 and 3.6). A number of experiments in Tab. 3.5 attain 13 as for robustness of the consequent processing of the experimental outputs the measurements in the central point [0,0] are repeated (4×).

In the discussed design of experiments (DoE), an introduction of the CCD (an arrangement of the input values) is interlaced with an applied RSM (the output values). This method is based on an introduction of the quadratic surface over the CCD nodes in the form

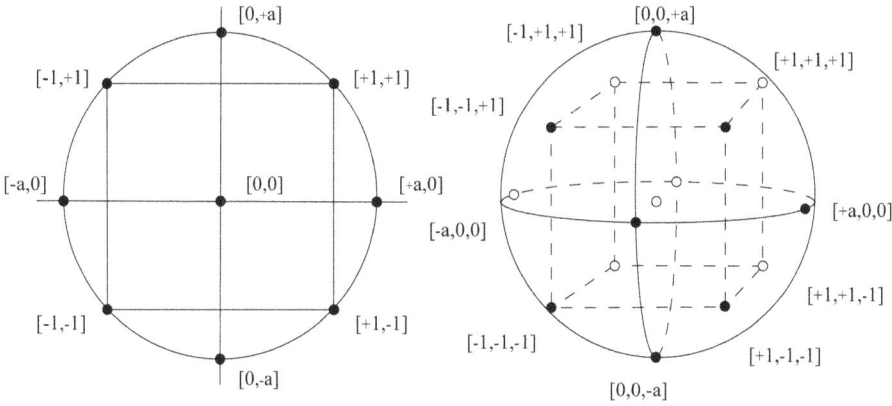

Fig. 3.14: Distribution of the experimental points for two (left) and three (right) factors.

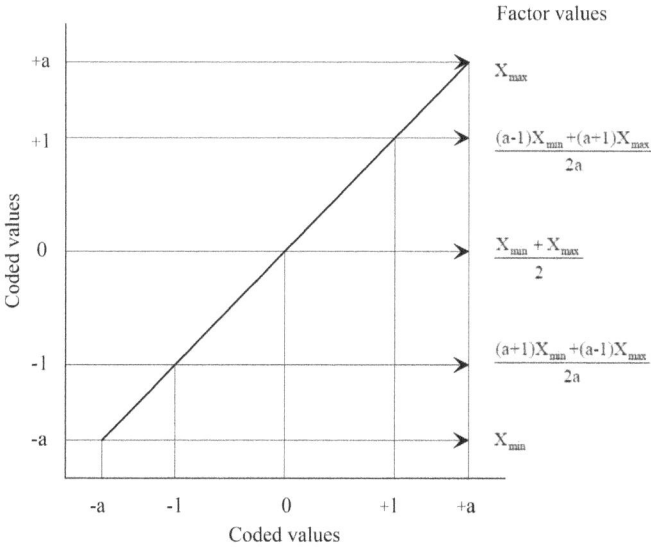

Fig. 3.15: A correspondence among the coded and factor values.

$$Y(X_1, ..., X_n) = C_0 + \sum_{i=1}^{n} C_i X_i + \sum_{i=1}^{n} C_{ii} X_i^2 + \sum_{i=1}^{n-1} \sum_{j=i+1}^{n} C_{ij} X_i X_j + \varepsilon, \qquad (3.2)$$

where Y is the predicted response (e.g. efficiency of turbidity or DOC removal), the parameter C_0 is the offset, the summation of terms with the parameters C_i represents a plane, the summation of terms with the parameters C_{ii} and C_{ij} represent the quadric surfaces; moreover the terms with the parameters C_{ij} describe the mutual interactions, and ε represents the deviation between the predicted and measured values.

Tab. 3.4: The actual factor values represented by the coded ones.

Coded value	$-a$	-1	0	$+1$	$+a$
Coagulant dose CD	CD_{min}	$\dfrac{(a+1)CD_{min} + (a-1)CD_{max}}{2a}$	$\dfrac{CD_{min}+CD_{max}}{2}$	$\dfrac{(a-1)CD_{min} + (a+1)CD_{max}}{2a}$	CD_{max}
pH	pH_{min}	$\dfrac{(a+1)pH_{min} + (a-1)pH_{max}}{2a}$	$\dfrac{pH_{min}+pH_{max}}{2}$	$\dfrac{(a-1)pH_{min} + (a+1)pH_{max}}{2a}$	pH_{max}
Flocculant dose FD	FD_{min}	$\dfrac{(a+1)FD_{min} + (a-1)FD_{max}}{2a}$	$\dfrac{FD_{min}+FD_{max}}{2}$	$\dfrac{(a-1)FD_{min} + (a+1)FD_{max}}{2a}$	FD_{max}

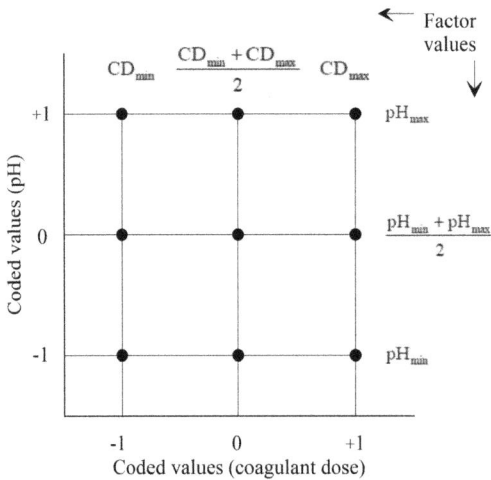

Fig. 3.16: A simple arrangement among the factor and coded values in the case of two factors.

Tab. 3.5: The actual factor values represented by the coded ones.

Coded value	-1	0	$+1$
Coagulant dose CD	CD_{min}	$\dfrac{CD_{min}+CD_{max}}{2}$	CD_{max}
pH	pH_{min}	$\dfrac{pH_{min}+pH_{max}}{2}$	pH_{max}

It is improbable that the second-order polynomial model can responsibly cover the whole region of relatively distant initial input variables but if these variables are reasonably restricted, then second-order approximation can be acceptable as it corresponds to the first three terms (groups – absolute, linear, and quadratic) in a Taylor expansion of more complicated functions. The parameters C_0, C_i, C_{ii}, and C_{ij} can be

Tab. 3.6: In total 13 combinations (jar-test experiments) for the 2D square arrangement.

Factors	Combinations of coded values												
	Run No. 1	Run No. 2	Run No. 3	Run No. 4	Run No. 5	Run No. 6	Run No. 7	Run No. 8	Run No. 9	Run No. 10	Run No. 11	Run No. 12	Run No. 13
Coagulant dose	−1	−1	−1	0	0	0	0	0	0	0	1	1	1
pH	−1	0	1	−1	0	0	0	0	0	1	−1	0	1

determined using a classical matrix-based linear regression analysis or using a specialised commercial software (both using the method of least squares).

The applied model (3.2) should exhibit consistency and stable variancy of the predicted response. In the case of the second-order polynomial model it should be reflected in its rotatability (Box and Hunter 1957), which is constant on spherical surfaces. This condition is fulfilled if $a = k^{1/2}$, where k is the number of factors. Hence, for two input variables $a \approx 1.414$ and for three input variables $a \approx 1.732$, see spherical CCD in Fig. 3.14. For these choices of a, the corresponding combinations of coded values are introduced in Tab. 3.7 (for $k = 2$ and $a \approx 1.414$), Tab. 3.8, and Fig. 3.17 (for $k = 3$ and $a \approx 1.732$). However, a choice of the value a is not unconditionally subject to the relation $a = k^{1/2}$ and also other values can prove to be useful in the specific cases.

Tab. 3.7: Combinations of coded values (with four repeated experiments in the central point) for the case of circumferential CCD.

Factors	Combinations of coded values												
	Run No. 1	Run No. 2	Run No. 3	Run No. 4	Run No. 5	Run No. 6	Run No. 7	Run No. 8	Run No. 9	Run No. 10	Run No. 11	Run No. 12	Run No. 13
Coagulant dose	−a	−1	−1	0	0	0	0	0	0	0	1	1	a
pH	0	−1	1	−a	0	0	0	0	0	a	−1	1	0

The above-introduced RSM in a combination with the CCD has been successfully applied to an optimisation of input variables in a coagulation/flocculation process in drinking water treatment. Trinh and Kang (2011) analysed removal efficiency of turbidity and DOC using aluminium sulphate (alum) and PACl using a schedule in Tab. 3.7. Zainal-Abideen et al. (2012) optimised turbidity removal, settling pH, and residual aluminium based on a combination of aluminium sulphate dosage, initial pH, and polymer dosage using an allocation in Tab. 3.8.

Tab. 3.8: Combinations of coded values (with four repeated experiments in the central point) for the case of spherical CCD.

Factors	Combinations of coded values																	
	Run No. 1	Run No. 2	Run No. 3	Run No. 4	Run No. 5	Run No. 6	Run No. 7	Run No. 8	Run No. 9	Run No. 10	Run No. 11	Run No. 12	Run No. 13	Run No. 14	Run No. 15	Run No. 16	Run No. 17	Run No. 18
Coagulant dose	$-a$	-1	-1	-1	-1	0	0	0	0	0	0	0	0	1	1	1	1	a
pH	0	-1	-1	1	1	$-a$	0	0	0	0	0	0	a	-1	-1	1	1	0
Flocculant dose	0	-1	1	-1	1	0	$-a$	0	0	0	0	a	0	-1	1	-1	1	0

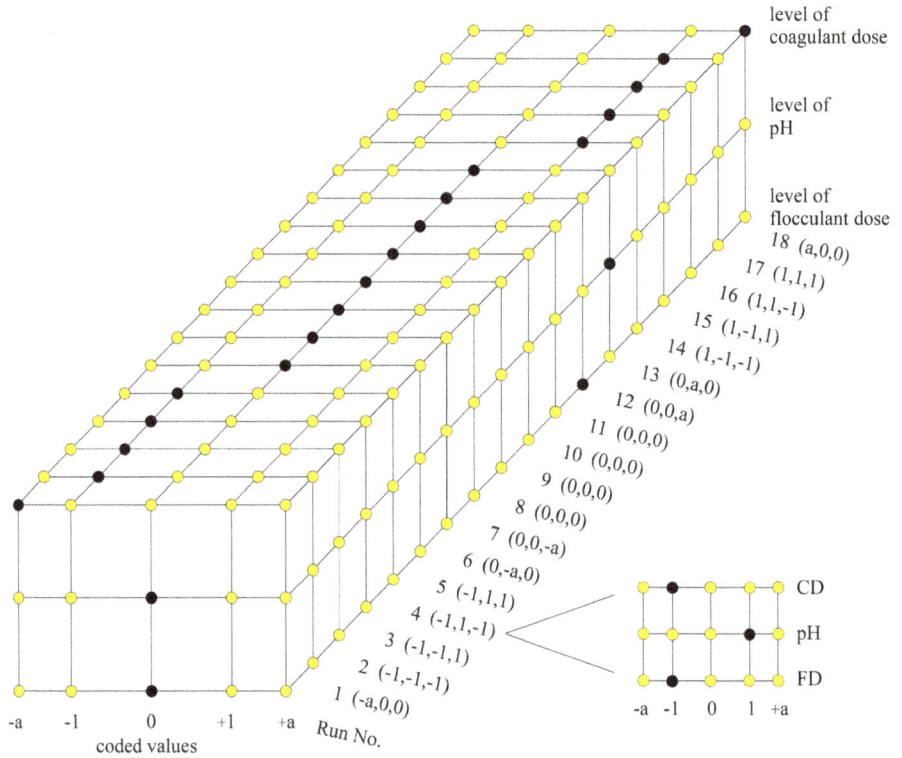

Fig. 3.17: In total 18 experimental measurements for three factors combing the input variables according to the spherical central composite design.

3.8 Impact of societal development on water treatment procedures

The industrialisation, climate changes, and other global factors significantly penetrate not only to water quantity but also radically alter water quality due to the presence of new contaminants. On the other hand, a development of new sophisticated experimental devices helps to discover harmful components that have not been known before. Hence, in tight cooperation with advances in medicine, it points not only to new pressing issues but also simultaneously unveil the hitherto unknown items which have to be addressed and are of primary concern.

3.8.1 Algal organic matters (AOMs)

Apart from the all-year-round contaminants there are also the seasonal ones where algae and cyanobacteria blooms are the main representatives. Their occurrence (and especially amount) is interlaced with the climate changes and increasing eutrophication. Their products cumulatively denoted as algal organic matters (AOMs) exhibit a series of unwanted phenomena as a presence of harmful cyanotoxins, the compounds deteriorating organoleptic properties (taste, odour) or containing sulphur, a contribution to potential bacterial growth in distribution systems. AOM also belongs to the precursors initiating a formation of DBP. The dominant components of AOM are saccharides (mono-, oligo-, and poly-saccharides) and nitrogenous substances (amino and nucleic acids, peptides, and proteins).

The crucial problem with AOM is given by its release into raw water in two forms: metabolic by-products denoted as extracellular organic matter and by means of cell lysis as cellular organic matter. Cell lysis is evoked by aging of algal population and aggressive physical and chemical processes (e.g. pre-oxidation). Cellular organic matter substantially worsens the coagulation/flocculation process as such, but also hinders the coagulation of other substances and also reduces efficacy of adsorption technique. Hence, AOMs influence not only a choice of coagulants/flocculants but also their dosage. Handling a specific AOM does not ensure general efficacy of water treatment with respect to AOM as AOM are formed by various phytoplankton species. It is not possible to expect that chemical composition of AOM in the specific water reservoirs will be repeated every season. It implies that the WTPs have to optimise their approaches every year.

The views to the topic of AOM from different approaches can be found in Henderson et al. (2008a), Li et al. (2012), Pivokonsky et al. (2014), Gonzalez-Torres et al. (2014), Pivokonsky et al. (2016), Hua et al. (2019), and Pivokonsky et al. (2021).

3.8.2 Pre-oxidation processes

A removal of AOM cannot be achieved by traditional water treatment techniques what resulted in an introduction of pre-oxidation process. The problem is that the applied chemical oxidants are not selective, i.e. apart from beneficial prevention from microbial replication (and hence causing effective inactivation of microorganisms) they also react with the dissolved constituents and particulates, which causes formation of harmful DBP. Ozone proved to be a very efficient oxidiser; however, at the presence of bromide the unacceptable brominated DBPs are formed as bromate BrO_3^-.

There are a series of classical pre-oxidation techniques applied to algal-rich raw water (the indicated references (reviews) summarise rich literature on the given topic):
– chlor(am)ination as discussed below (Dong et al. 2021);
– pre-ozonation (Jiang et al. 2019);
– ferrate(VI) treatment (Jiang et al. 2016);
– potassium permanganate $KMnO_4$ oxidation (Henderson 2008b; Naceradska et al. 2017).

Simultaneously there are applied the so-called advanced oxidation processes combining oxidants, radiation, and catalysts (hence chemical and purely physical approaches) as for instance O_3/H_2O_2, UV/H_2O_2, UV/O3, UV/TiO_2, Fe^{2+}/H_2O_2, vacuum ultraviolet radiation, and ionising radiation (Matilainen and Sillanpää 2010; Gil et al. 2019, Lei et al. 2021; Li et al. 2023).

3.8.3 Chlor(am)ination and disinfection by-products (DBP)

The most traditional disinfection procedures removing ubiquitous pathogenic (infectious) microorganisms (such as bacteria, viruses, and protozoa) from drinking water use chlorine, chloramines, and chlorine dioxide. Classically used chlorine is gradually replaced by chloramine (strictly said monochloramine, see Fig. 3.18) which is – in contrast to chlorine – more acceptable in two aspects: lower reactiveness and lower dissipation.

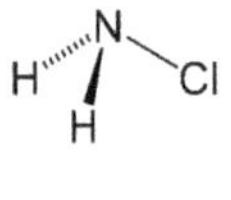

Fig. 3.18: Chemical structure of chloramine (monochloramine).

However, reaction of disinfectants with organic and inorganic matter initiates forming of the so-called DBP with an adverse impact on human health. Bellar et al. (1974) were the first who recognised this reverse side of disinfection procedure. An application of chlorine causes a formation of a series of DBPs including trihalomethanes (THM) and haloacetic acids (HAA). THMs are chemical compounds where three out of four hydrogen atoms of methane (CH₄) are replaced by halogen atoms. The Environmental Protection Agency (EPA) singled out four constituents (chloroform, bromoform, bromodichloromethane, and dibromochloromethane, see Fig. 3.19) denoted as the total trihalomethanes (TTHM) which presence should not exceed 80 parts per billion in treated water. HAAs are carboxylic acids where one or more hydrogen atoms in the methyl group of acetic acid are replaced by halogen atoms. As the HAAs are highly chemically stable, they persist in water after formation. The five HAAs with the most frequent occurrence in water (chloroacetic acid, dichloroacetic acid, trichloroacetic acid, bromoacetic acid, and dibromoacetic acid) are denoted as HAA5 (see Fig. 3.20) and are regulated by the EPA. This list is also widened to HAA9 with the corresponding frequency. The analogous measures are also taken in the Directive (EU) 2020/2184.

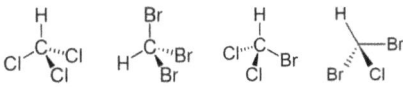

Fig. 3.19: Chemical structures of the total trihalomethanes (from left): chloroform, bromoform, bromodichloromethane, and dibromochloromethane.

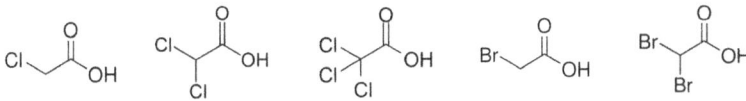

Fig. 3.20: Chemical structures of HAA5 (from left): chloroacetic acid, dichloroacetic acid, trichloroacetic acid, bromoacetic acid, and dibromoacetic acid.

Reaction of chloramine with the organic and inorganic (e.g. bromide) matter produces much lower presence of THM and HAA in drinking water. However, this seemingly better tendency to make water healthier is torpedoed – in the case of chloramine – by forming of dangerous *N*-nitrosodimethylamine (NDMA) (see Fig. 3.21). This compound – also known as dimethylnitrosamine – is highly hepatotoxic and carcinogenic, more detailed analysis is presented in Krasner et al. (2013). NDMA occurrence in drinking water is regulated up to single nanograms per litre.

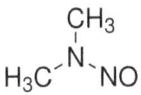

Fig. 3.21: Chemical structure of *N*-nitrosodimethylamine.

The known number of DBP is strikingly increasing. While Sadiq and Rodriguez (2004) introduce that "More than 250 DBPs have been identified, but the behavioural profile of only approximately 20 DBPs are adequately known.", Richardson (2011) reports over 600 characterised DBP, and Richardson and Kimura (2020) already mention over 700 DBPs

with a commentary that most organic halogen products resulting from chlorination re-
main unidentified. Moreover, with further research and new experimental techniques
it is apparent that certain emerging DBPs are substantially (orders of magnitude) more
toxic than the currently regulated DBPs (Richardson and Plewa 2020; Krasner et al.
2022). However, their occurrence due to their concentration is not always easily detect-
able (Li and Mitch 2018; Wawryk et al. 2021). The emerging DBPs with risks to human
health usually cannot be removed by conventional water treatment.

A possible reduction of the DBP level can be related to a description of their precursors with the aim of their suppression. Historically, NOM is taken as a principal
precursor. Nevertheless, from the viewpoint of NDMA presence more challenging are
AOM and treated wastewater effluent organic matter rich in nitrogen (Krasner et al.
2012). There are also other precursors such as pesticides, pharmaceuticals, bromide,
and iodide (Ding et al. 2019; Gilca et al. 2020). It is apparent that the best efficiency in
reducing DBPs presence can be achieved by maximal elimination of DBP precursors,
in this context to consider a possibility of mutual successiveness of an application of
coagulant/flocculant agents and the process of disinfection.

3.9 Mechanical aspects of aggregation

A jar test represents a very effective method how to choose a suitable coagulant/floc-
culant, in which dosage and under which pH number. However, it is necessary to take
into account that an arrangement of this device is optimised from the hydrodynamic
viewpoint. The studied batches (not exceeding 2 L in volume) are practically immediately homogenised and the input values of the above-stated quantities correspond to
an idealised conditions.

It is possible to say that this approach only partially imitate the reality in the
drinking water treatment plants (DWTPs). The crucial difference is a degree of homogenisation. While the arrangement of jar tests practically ensures complete homogenisation, the reality in DWTPs diametrically differs, and among other things, there
can arise the problems with appearance of problematic zones.

Roughly said there are two possibilities how to minimise a deviation from 'jar test'
conditions based on mechanical measures. The static (hydraulic) and dynamic (stirring)
measures differ in the financial costs. While the first ones are covered nearly exclusively with their acquisition costs, the second ones (stirrers) consume electric energy,
require maintenance costs, and have to be occasionally (regularly) replaced.

The classical static perforated walls contribute to better homogenisation of treated
water and intensity of shearing in dependence on the discharge of water and geometry
of the hole rims. The geometrical arrangements of the sedimentation tanks should eliminate problematic zones and gradually reduce shearing of the formed aggregates and
thus prevent from their breakage at the end of the treatment process, which is achieved
for instance by gradual prolongation of the individual compartments (see Fig. 3.22).

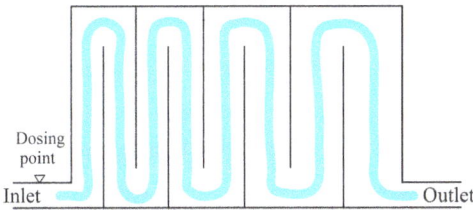

Fig. 3.22: Gradual prolongation of the consecutive compartments in the sedimentation tank contributing to shear reduction and compactness of aggregates.

The stirrers represent an integral part of DWTPs as they substantially improve homogenisation of coagulant/flocculant into treated water. They perform two roles: the positive way is an improvement of distribution of agents within the treated bulk, the negative one can contribute to breakage of already created clusters. Nevertheless, the principal aim is to maintain a turbulent flow regime. This requires a sufficiently high power surpassing the resistance of water against flow. There is a series of mechanical stirrers (Bratby 2016), for instance, the hyperboloid mixers (see Fig. 3.23) range to the most popular in the recent period (Invent 2025).

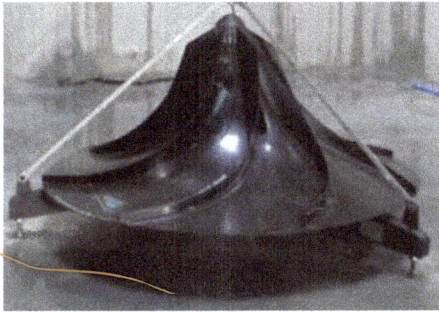

Fig. 3.23: The hyperboloid mixer (https://invent-uv.com/products-services/mixing-technology/hyperdive-mixer/).

As pressure p exerted to fluid by the impeller is proportional to fluid density ρ and the square of velocity ($\rho \propto \rho v^2$), and velocity is proportional to rotational speed n and the impeller diameter D ($v \propto n.D$), the pressure is given by

$$p \propto \rho n^2 D^2. \tag{3.3}$$

The power P is expressed as the product of the velocity v and the force F ($P = v.F$), which can be determined from the relation $p = F/A$. Here A represents an area on which the force exerts and which is proportional to D^2. Summarising these relations we obtain

$$P \propto \rho n^3 D^5. \tag{3.4}$$

The dimensionless power number N_P is thus defined as

$$N_P = \frac{P}{\rho n^3 D^5} \tag{3.5}$$

relating the resistance force to the inertia force and indicating which force prevails or dominates.

Another expression for P follows from a rheological description of flow conditions (see Fig. 3.24). The shear stress τ_{shear} can be expressed in two ways: either using a hydrodynamic relation F/A (as denoted in Fig. 3.24) or using a relation based on rheological considerations with a dynamic viscosity μ. Comparing both relations

$$\frac{F}{A} = \mu \left(\frac{dv}{dy}\right)_{y=0} \tag{3.6}$$

we obtain

$$F = A\mu G, \tag{3.7}$$

where – a time-derivative of shear strain γ – G is the shear rate (velocity gradient, $G = d\gamma/dt$). The shear rate (velocity gradient) expresses the rate of change in velocity at which one layer passes over the adjacent one. From here we obtain a relation for the rate of dissipation P following the definition as the product of force F and velocity Gh (representing a difference in velocity, see Fig. 3.24)

$$P = (Gh) \times (A\mu G). \tag{3.8}$$

Denoting a volume $V = h \times A$, we get the well-known relation for the velocity gradient

$$G = \sqrt{\frac{P}{\mu V}}. \tag{3.9}$$

The velocity gradient G represents one of the most traditionally considered parameters for describing the process of flocculation because it is considered as an indicator in evaluating the degree of mixing imparted to the treated water and hence, formation of the flocs and their growth. Mutl et al. (2006) show that the aggregates formed under higher velocity gradient G are more compact and homogeneous in size and resistant to breakage. On the other hand, "G factor does not consider how effectively a particular mixer design converts power to mixture quality, neither does it consider how effective the injection method is" (Statiflo 2025).

The compactness of flocs is a key factor what concerns a potential floc breakage and sedimentation. A measure of compactness – expressed by means of a fractal dimension – is discussed in the following chapter.

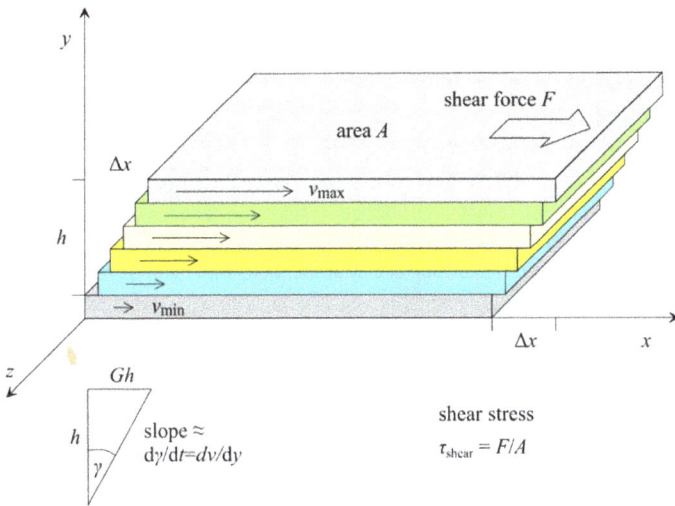

Fig. 3.24: Shearing of the individual layers along the flow direction.

References

Abu Hasan, H., Muhammad, M.H., Ismail, N.I. (2020) A review of biological drinking water treatment technologies for contaminants removal from polluted water resources. Journal of Water Process Engineering 33, 101035. https://doi.org/10.1016/j.jwpe.2019.101035

Adusei-Gyamfi, J., Ouddane, B., Rietveld, L., Cornard, J.-P., Criquet, J. (2019) Natural organic matter-cations complexation and its impact on water treatment: A critical review. Water Research 160, 130–147. https://doi.org/10.1016/j.watres.2019.05.064

Anderson, L.E., DeMont, I., Dunnington, D.D., Bjorndahl, P., Redden, D.J., Brophy, M.J., Gagnon, G.A. (2023) A review of long-term change in surface water natural organic matter concentration in the northern hemisphere and the implications for drinking water treatment. Science of the Total Environment 858, 159699. https://doi.org/10.1016/j.scitotenv.2022.159699

Bellar, T.A., Lichtenberg, J.J., Kroner, R.C. (1974) The occurrence of organohalides in chlorinated drinking waters. Journal of the American Water Works Association 66, 703–706. https://doi.org/10.1002/j.1551-8833.1974.tb02129.x

Bhatnagar, A.; Sillanpää, M. (2017) Removal of natural organic matter (NOM) and its constituents from water by adsorption – A review. Chemosphere 166, 497–510. http://dx.doi.org/10.1016/j.chemo sphere.2016.09.098

Box, G.E.P., Hunter, J.S. (1957) Multi-factor experimental designs for exploring response surfaces. Annals of Mathematical Statistics 28, 195–241. http://dx.doi.org/10.1214/aoms/1177707047

Bratby, J. (2016) Coagulation and Flocculation in Water and Wastewater Treatment. 3rd ed., IWA Publishing.

Chen, W., Yu, H.Q. (2021) Advances in the characterization and monitoring of natural organic matter using spectroscopic approaches. Water Research 190, 116759. https://doi.org/10.1016/j.watres.2020.116759

Ciobanu, R., Mihai, M., Teodosiu, C. (2022) An overview of natural organic matter removal by coagulation in drinking water treatment. Buletinul Institutului Politehnic din Iasi 68, 69–92. https://www.re searchgate.net/publication/366808403

Davis, C.C., Edwards, M. (2014) Coagulation with hydrolyzing metal salts: Mechanisms and water quality impacts. Critical Reviews in Environmental Science and Technology 44, 303–347. https://doi.org/10.1080/10643389.2012.718947

Ding, S., Deng, Y., Bond, T., Fang, C., Cao, Z., Chu, W. (2019) Disinfection byproduct formation during drinking water treatment and distribution: A review of unintended effects of engineering agents and materials. Water Research 160, 313–329. https://doi.org/10.1016/j.watres.2019.05.024

Dong, F.L., Lin, Q.F., Li, C., He, G.L., Deng, Y. (2021) Impacts of pre-oxidation on the formation of disinfection byproducts from algal organic matter in subsequent chlor(am)ination: A review. Science of the Total Environment 754, 141955. https://doi.org/10.1016/j.scitotenv.2020.141955

Duan, J. (1997) Influence of Dissolved Silica on Flocculation of Clay Suspensions with Hydrolysing Metal Salts. PhD Thesis, University of London.

Duan, J., Gregory, J. (2003) Coagulation by hydrolyzing metal salts. Advances in Colloid and Interface Science 100–102, 475–502. https://doi.org/10.1016/S0001-8686(02)00067-2

Edzwald, J.K. (1993) Coagulation in drinking water treatment: Particles, organics and coagulants. Water Science & Technology 27, 21–35. https://doi.org/10.2166/wst.1993.0261

Gil, A., Galeano, L.A., Vicente M.Á. (eds.) (2019) Applications of Advanced Oxidation Processes (AOPs) in Drinking Water Treatment. Springer. https://doi.org/10.1007/978-3-319-76882-3

Gilca, A.F., Teodosiu, C., Fiore, S., Musteret, C.P. (2020) Emerging disinfection byproducts: A review on their occurrence and control in drinking water treatment processes. Chemosphere 259, 127476. https://doi.org/10.1016/j.chemosphere.2020.127476

Gonzalez-Torres, A., Putnam, J., Jefferson, B., Stuetz, R.M., Henderson, R.K. (2014) Examination of the physical properties of Microcystis aeruginosa flocs produced on coagulation with metal salts. Water Research 60, 197–209. https://doi.org/10.1016/j.watres.2014.04.046

Gregory, J. (2006) Particles in Water. Properties and Processes. Taylor & Francis Group, Boca Raton, Florida, USA.

Guldberg, C.M., Waage, P. (1879) Ueber die chemische Affinität. Journal für praktische Chemie 19, 69–114. https://doi.org/10.1002/prac.18790190111

Hanson, A.T., Cleasby, J.L. (1990) The effects of temperature on turbulent flocculation – fluid-dynamics and chemistry. Journal of the American Water Works Association 82, 56–73. https://doi.org/10.1002/j.1551-8833.1990.tb07053.x

Henderson, R., Parsons, S.A., Jefferson, B. (2008b) The impact of algal properties and peroxidation on solid–liquid separation of algae. Water Research 42, 1827–1845. https://doi.org/10.1016/j.watres.2007.11.039

Henderson, R., Sharp, E., Jarvis, P., Parsons, S., Jefferson, B. (2006) Identifying the linkage between particle characteristics and understanding coagulation performance. Water Science and Technology: Water Supply 6, 31–38. https://doi.org/10.2166/ws.2006.005

Henderson, R.K., Baker, A., Parsons, S.A., Jefferson, B. (2008a) Characterisation of algogenic organic matter extracted from cyanobacteria, green algae and diatoms. Water Research 42, 3435–3445. https://doi.org/10.1016/j.watres.2007.10.032

Hua, L.C., Lai, C.H., Wang, G.S., Lin, T.F., Huang, C. (2019) Algogenic organic matter derived DBPs: Precursor characterization, formation, and future perspectives – A review. Critical Reviews in Environmental Science and Technology 49, 1803–1834. https://doi.org/10.1080/10643389.2019.1586057

Hummel, W., Berner, U., Curti, E., Pearson, F.J., Thoenen, T. (2002) Nagra / PSI Chemical Thermodynamic Data Base 01/01, Technical Report 02–16 (p. 94). Radiochimica Acta 90, 805–813. https://doi.org/10.1524/ract.2002.90.9-11_2002.805

Invent (accessed 2025, Mar 27) https://invent-uv.com/products-services/mixing-technology/hyperdive-mixer/

Jacangelo, J.G., DeMarco, J., Owen, D.M., Randtke, S.J. (1995) Selected processes for removing NOM: an overview. Journal of the American Water Works Association 87, 64–77. https://doi.org/10.1002/j.1551-8833.1995.tb06302.x

Jarvis, P., Banks, J., Molinder, R., Stephenson, T., Parsons, S.A., Jefferson, B. (2008) Processes for enhanced NOM removal: beyond Fe and Al coagulation. Water Supply 8, 709–716. https://doi.org/10.2166/ws.2008.155

Jarvis, P., Jefferson, B., Parsons, SA. (2004) Characterising natural organic matter flocs. Water Supply 4, 79–87. https://doi.org/10.2166/ws.2004.0064

Jiang, Y.J., Goodwill, J.E., Tobiason, J.E., Reckhow, D.A. (2016) Impacts of ferrate oxidation on natural organic matter and disinfection byproduct precursors. Water Research 96, 114–125. https://doi.org/10.1016/j.watres.2016.03.052

Jiang, Y.J., Goodwill, J.E., Tobiason, J.E., Reckhow, D.A. (2019) Comparison of ferrate and ozone pre-oxidation on disinfection byproduct formation from chlorination and chioramination. Water Research 156, 110–124. https://doi.org/10.1016/j.watres.2019.02.051

Kang, L.S., Cleasby, J.L. (1995) Temperature effects on flocculation kinetics using Fe(III) coagulant. Journal of Environmental Engineering 121, 893–901. https://doi.org/10.1061/(ASCE)0733-9372(1995)121:12(893)

Khuri, A.I., Cornell, J.A. (2018) Response Surfaces: Designs and Analyses. 2nd ed., CRC Press, Boca Raton. https://doi.org/10.1201/9780203740774

Kosobucki, P., Buszewski, B. (2014) Natural organic matter in ecosystems – a review. Nova Biotechnologica et Chimica 13, 109–129. https://doi.org/10.1515/nbec-2015-0002

Krasner, S.W., Jia, A., Lee, C.-F.T., Shirkhani, R., Allen, J.M., Richardson, S.D., Plewa, M.J. (2022) Relationships between regulated DBPs and emerging DBPs of health concern in U.S. drinking water. Journal of Environmental Sciences 117, 161–172. https://doi.org/10.1016/j.jes.2022.04.016

Krasner, S.W., Mitch, W.A., McCurry, D.L., Hanigan, D., Westerhoff, P. (2013) Formation, precursors, control, and occurrence of nitrosamines in drinking water: a review. Water Research 47, 4433–4450. https://doi.org/10.1016/j.watres.2013.04.050

Krasner, S.W., Mitch, W.A., Westerhoff, P., Dotson, A. (2012) Formation and control of emerging C- and N-DBPs in drinking water. Journal of the American Water Works Association 104, E582–E595. https://doi.org/10.5942/jawwa.2012.104.0148

Lei, X., Lei, Y., Zhang, X.R., Yang, X. (2021) Treating disinfection byproducts with UV or solar irradiation and in UV advanced oxidation processes: A review. Journal of Hazardous Materials 408, 124435. https://doi.org/10.1016/j.jhazmat.2020.124435

Li, J., Zhang, Z., Xiang, Y.Y., Jiang, J., Yin, R. (2023) Role of UV-based advanced oxidation processes on NOM alteration and DBP formation in drinking water treatment: A state-of-the-art review. Chemosphere 311, 136870. https://doi.org/10.1016/j.chemosphere.2022.136870

Li, L., Gao, N.Y., Deng, Y., Yao, J.J., Zhang, K.J. (2012) Characterization of intracellular & extracellular algae organic matters (AOM) of *Microcystic aeruginosa* and formation of AOM-associated disinfection byproducts and odor & taste compounds. Water Research 46, 1233–1240. https://doi.org/10.1016/j.watres.2011.12.026

Li, X.-F., Mitch, W.A. (2018) Drinking water disinfection byproducts (DBPs) and human health effects: Multidisciplinary challenges and opportunities. Environmental Science & Technology 52, 1681–1689. https://doi.org/10.1021/acs.est.7b05440

Matilainen, A., Gjessing, E.T., Lahtinen, T., Hed, L., Bhatnagar, A., Sillanpää, M. (2011) An overview of the methods used in the characterisation of natural organic matter (NOM) in relation to drinking water treatment. Chemosphere 83, 1431–1442. https://doi.org/10.1016/j.chemosphere.2011.01.018

Matilainen, A., Lindqvist, N., Tuhkanen, T. (2005) Comparison of the efficiency of aluminium and ferric sulphate in the removal of natural organic matter during drinking water treatment process. Environmental Technology 26, 867–875. https://doi.org/10.1080/09593332608618502

Matilainen, A., Sillanpää, M. (2010) Removal of natural organic matter from drinking water by advanced oxidation processes. Chemosphere 80, 351–365. https://doi.org/10.1016/j.chemosphere.2010.04.067

Matilainen, A., Vepsäläinen, M., Sillanpää, M. (2010) Natural organic matter removal by coagulation during drinking water treatment: A review. Advances in Colloid and Interface Science 159, 189–197. https://doi.org/10.1016/j.cis.2010.06.007

Menya, E., Olupot, P.W., Storz, H., Lubwama, M., Kiros, Y. (2018) Production and performance of activated carbon from rice husks for removal of natural organic matter from water: A review. Chemical Engineering Research and Design 129, 271–296. https://doi.org/10.1016/j.cherd.2017.11.008

Montgomery, D.C. (2013) Design and Analysis of Experiments. 8th ed., Wiley.

Mutl, S., Polasek, P., Pivokonsky, M., Kloucek, O. (2006) The influence of G and T on the course of aggregation in treatment of medium polluted surface water. Water Supply 6, 39–48. https://doi.org/10.2166/ws.2006.006

Naceradska, J., Pivokonsky, M., Pivokonska, L., Baresova, M., Henderson, R.K., Zamyadi, A., Janda, V. (2017) The impact of pre-oxidation with potassium permanganate on cyanobacterial organic matter removal by coagulation. Water Research 114, 42–49. https://doi.org/10.1016/j.watres.2017.02.029

Pan, Y., Li, H., Zhang, X.R., Li, A.M. (2016) Characterization of natural organic matter in drinking water: Sample preparation and analytical approaches. Trends in Environmental Analytical Chemistry 12, 23–30. https://doi.org/10.1016/j.teac.2016.11.002

Pivokonsky, M., Kopecka, I., Cermakova, L., Fialova, K., Novotna, K., Cajthaml, T., Henderson, R.K., Pivokonska, L. (2021) Current knowledge in the field of algal organic matter adsorption onto activated carbon in drinking water treatment. Science of the Total Environment 799, 149455. https://doi.org/10.1016/j.scitotenv.2021.149455

Pivokonsky, M., Naceradska, J., Kopecka, I., Baresova, M., Jefferson, B., Li, X., Henderson, R.K. (2016) The impact of algogenic organic matter on water treatment plant operation and water quality: a review. Critical Reviews in Environmental Science and Technology 46, 291–335. https://doi.org/10.1080/10643389.2015.1087369

Pivokonsky, M., Novotna, K., Petricek, R., Cermakova, L., Prokopova, M., Naceradska, J. (2024) Fundamental chemical aspects of coagulation in drinking water treatment – Back to basics. Journal of Water Process Engineering 57, 104660. https://doi.org/10.1016/j.jwpe.2023.104660

Pivokonsky, M., Safarikova, J., Baresova, M., Pivokonska, L., Kopecka, I. (2014) A comparison of the character of algal extracellular versus cellular organic matter produced by cyanobacterium, diatom and green alga. Water Research 51, 37–46. https://doi.org/10.1016/j.watres.2013.12.022

Richardson, S.D. (2011) Disinfection by-products: formation and occurrence in drinking water. In Encyclopedia of Environmental Health, ed. Nriagu, J.O., Elsevier, Inc. Press, pp. 110–136.

Richardson, S.D., Kimura, S.Y. (2020) Water analysis: emerging contaminants and current issues. Analytical Chemistry 92, 473–505. https://doi.org/10.1021/acs.analchem.9b05269

Richardson, S.D., Plewa, M.J. (2020) To regulate or not to regulate? What to do with more toxic DBPs. Journal of Environmental Chemical Engineering 8, 103939. https://doi.org/10.1016/j.jece.2020.103939

Riyadh, A., Peleato, N.M. (2024) Natural organic matter character in drinking water distribution systems: A review of impacts on water quality and characterization techniques. Water 16, 446. https://doi.org/10.3390/w16030446

Sadiq, R., Rodriguez, M.J. (2004) Disinfection by-products (DBPs) in drinking water and predictive models for their occurrence: a review. Science of the Total Environment 321, 21–46. https://doi.org/10.1016/j.scitotenv.2003.05.001

Sharp, E.L., Parsons, S.A., Jefferson, B. (2006) Seasonal variations in natural organic matter and its impact on coagulation in water treatment. Science of the Total Environment 363, 183–194. https://doi.org/10.1016/j.scitotenv.2005.05.032

Sillanpää, M., Ncibi, M.C., Matilainen, A., Vepsäläinen, M. (2018) Removal of natural organic matter in drinking water treatment by coagulation: A comprehensive review. Chemosphere 190, 54–71. https://doi.org/10.1016/j.chemosphere.2017.09.113

Sørensen, S.P.L. (1909) Über die Messung und die Bedeutung der Wasserstoffionenkonzentration bei enzymatischen Prozessen. Biochemische Zeitschrift 21, 131–304. https://publikationen.ub.uni-frankfurt.de/frontdoor/index/index/year/2007/docId/17417

Statiflo (accessed 2025, Mar 27) https://statiflo.com/about-static-mixing-2/mixture-quality/g-factor-velocity-gradient/

Stefansson, A. (2007) Iron(III) hydrolysis and solubility at 25 °C. Environmental Science & Technology 41, 6117–6123. https://doi.org/10.1021/es070174h

Trinh, T.K., Kang, L.S. (2011) Response surface methodological approach to optimize the coagulation–flocculation process in drinking water treatment. Chemical Engineering Research and Design 89, 1126–1135. https://doi.org/10.1016/j.cherd.2010.12.004

Van Benschoten, J.E., Edzwald, J.K. (1990) Chemical aspects of coagulation using aluminum salts – II. coagulation of fulvic-acid using alum and polyaluminum chloride. Water Research 24, 1527–1535. https://doi.org/10.1016/0043-1354(90)90087-M

Waage, P., Guldberg, C.M. (1864) Studier over affiniteten. Forhandlinger i Videnskabs-selskabet i Christiania 1, 35–45.

Wawryk, N.J.P., Craven, C.B., Blackstock, L.K.J., Li, X.F. (2021) New methods for identification of disinfection byproducts of toxicological relevance: Progress and future directions. Journal of Environmental Sciences 99, 151–159. https://doi.org/10.1016/j.jes.2020.06.020

Zainal-Abideen, M., Aris, A., Yusof, F., Abdul-Majid, Z., Selamat, A., Omar, S.I. (2012) Optimizing the coagulation process in a drinking water treatment plant – comparison between traditional and statistical experimental design jar tests. Water Science & Technology 65, 496–503. https://doi.org/10.2166/wst.2012.561

Zhang, Y., Zhao, X.H., Zhang, X.B., Peng, S. (2015) A review of different drinking water treatments for natural organic matter removal. Water Supply 15, 442–455. https://doi.org/10.2166/ws.2015.011

Chapter 4
Fractal nature of aggregates, experimental determination

4.1 Introduction to a fractal dimension

The classical Euclidean geometry grouping all objects to four discrete categories is relatively strict as all distinct objects are classified in one out of only four groups:
- 0-dimensional (0D) objects: points,
- 1-dimensional (1D) objects: straight lines,
- 2-dimensional (2D) objects: planar objects,
- 3-dimensional (3D) objects: spatial objects.

This is interlaced with the expression for the determination of their basic characteristics (bc): length, area, and volume:

$$\mathrm{bc} \sim \mathrm{Const} \times \ell^{D}, \tag{4.1}$$

where l represents the length (diameter of a circle, edge of a cube) and D is the corresponding integer (dimension).

   This classical – discrete – classification distributes all objects into three (if we exclude the most trivial objects: points) groups. This seems to be really rough differentiation because it cumulates apparently different objects – such as a solid sphere and a fluffy porous prism – into one group, specifically 3D objects. The opposite example can be found in the infinitesimal calculus concerning a computation of the definite integrals expressing the signed area (positive and negative increments specified by the positive or negative sign – 'above or below an axis'). In this case it seems natural that an area between the axis and the function should be preserved if a limited number of points on the curve are omitted.

   Both cases (better classification (diversification) of the objects and more intuitive estimate of an area) can be generalised by introducing the so-called measure theory (Halmos 1974) which eliminates the problems with the discrete concept – either a restriction only to the integer dimensions or a dependence on countable set of points in evaluating the definite integrals.

   Let us start very shortly with the definite integrals because it is connected with better intuition. The common integrals as

$$\int_{0}^{\pi} \sin(x)dx = 2 \tag{4.2}$$

https://doi.org/10.1515/9783111246765-004

are of the Newton-type supposing continuity of an integrand (sin(x)). This definite integral expresses an area between the axis x and the sine curve (=2). Intuitively, this result can also be expected if a couple of points are eliminated from a course of the sine function (expressed by the points in Fig. 4.1). In this case the adapted function is no longer continuous, and hence, no primitive function exists. Using the so-called Lebesgue measure the validity of the result (area = 2) can be generalised for 'perforated' sine functions (with the eliminated points). In the sense of Lebesgue approach the 'weight' of perforated x-interval between 0 and π is still π and even in the case that all rational numbers are excluded from the interval $[0,\pi]$ as the rational numbers represent a countable set (a set is countable if either a number of its elements is finite or the elements of the set can be enumerated by integers – one-to-one correspondence).

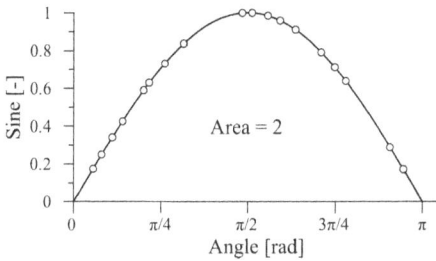

Fig. 4.1: Integration of discontinuous functions.

Another approach – parallel to the Lebesgue measure – is represented by the Hausdorff measure which – in contrast to the traditional Euclidean notions of length, area, and volume – enables (through the so-called Hausdorff dimension) to ascribe to the studied objects also non-integer dimensions starting from 0 and ending with 3. The related notion of fractal dimension was popularised very much especially in the 80s (starting with Mandelbrot 1967, who himself coined a term 'fractal'). The term 'fractal dimension' can be expressed and documented on the examples more easily than the term 'Hausdorff measure' based on countable coverage of a set.

Mandelbrot presented the term fractal dimension on the example of the 'coastline paradox'. The measured length of a coastline non-negligibly depends on the length of the used 'ruler'. With the decreasing ruler length, the coastline length is still increasing. In the case of the Great Britain the couples (length of a ruler, length of a coastline) develop in the following way (in km): (100, 2800) and (50, 3400). With a decreasing length of the ruler, a coastline length tends to infinity (see Fig. 4.2).

This result concerns every island and makes no difference among them. However, by suitable processing of the individual steps it is possible to obtain a characteristic – fractal dimension – reflecting roughness (irregularity) of the individual coastlines. If we denote a length scale (a ruler length) by ε and a number of ruler frequency along an island perimeter by N (the length of the approximated perimeter equals $N \times \varepsilon$), the dependence of $N(\varepsilon)$ on ε

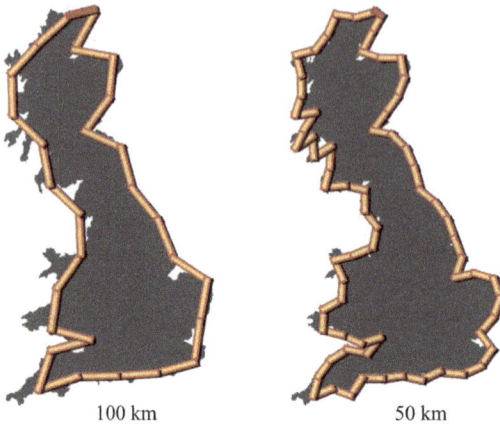

100 km   50 km

Fig. 4.2: Approximation of the length of a UK coastline in dependence on a ruler length (taken from https://en.wikipedia.org/wiki/Coastline_paradox).

$$N(\varepsilon) = \text{Const} \times \varepsilon^{-D}, \tag{4.3}$$

can be depicted in the log-log coordinates (see Fig. 4.3), where the couple of experimental points are introduced above. Not only these points lie on a straight line but approximately all the experimental points. This finding holds for any island ($D = 1.25$ (UK), 1.13 (Australia), and 1.52 (Norway)). The parameter D denotes a fractal dimension, and its value – as expected – is between 1 and 2 and describes differences in the individual coastlines.

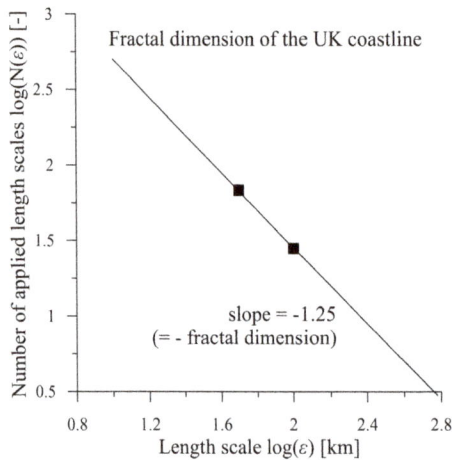

Fig. 4.3: Linear dependence of $N(\varepsilon)$ on a scale ε in the log-log coordinates.

If we express a length of the coastline in the classical (1D) Euclidean dimension

$$\ell(\varepsilon) = N(\varepsilon) \times \varepsilon = \text{Const} \times \varepsilon^{-D} \times \varepsilon = \text{Const} \times \varepsilon^{1-D}, \qquad (4.4)$$

then the length ℓ tends to infinity. However, if we determine a length of the coastline in the space of dimension D,

$$\ell(\varepsilon) = N(\varepsilon) \times \varepsilon^{D} = \text{Const} \times \varepsilon^{-D} \times \varepsilon^{D} = \text{Const}, \qquad (4.5)$$

we see that the length is finite and a space of the dimension D is the only space, where the length exhibits a finite positive number.

A term 'fractal' – as introduced by Mandelbrot – denotes a geometrical shape with detailed structure at arbitrarily small level which is successively reproducing its form. In other words, a defining property of the fractal is self-similarity, which refers to an infinite nesting of structure on all scales. Strict self-similarity refers to a characteristic of the form exhibited when a substructure resembles a superstructure in the same form.

This notion can be illustrated by the following three classical examples consecutively representing geometrical objects with a fractal dimension between 0 and 1, 1 and 2, and 2 and 3. Their constructions are connected with the starting shape (initiator) and the derived one (generator). Then this procedure proceeds step by step.

The Cantor set is initiated by a linear segment followed at level 1 by removing the inner third (open interval). In any other level the inner thirds (open intervals) of the preceding segments are removed (see Fig. 4.4).

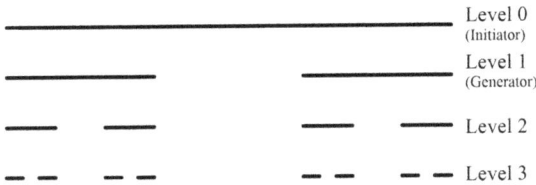

Fig. 4.4: The Cantor set, fractal dimension $D \cong 0.63$.

From the construction of the Cantor set it follows that a length of the successive level is two thirds of the preceding one with a scale one third (a length of successive subsegments equals one third of the preceding one)

$$\ell(\varepsilon/3) = 2/3 \times \ell(\varepsilon). \qquad (4.6)$$

Using eq. (4.4)

$$\text{Const} \times (\varepsilon/3)^{1-D} = (2/3) \times \text{Const} \times \varepsilon^{1-D}, \qquad (4.7)$$

we obtain

$$D = \ln 2 / \ln 3 \cong 0.63. \tag{4.8}$$

The Cantor set is an example of passage from a curve (linear segment) with classical 1D topology to a set of the fractal dimension lower than one (no infinitely small continuous segment can be contained in the Cantor set; otherwise the dimension would equal one). The opposite example – a passage from 1D curve (linear segment) to an object with fractal dimension exceeding one – is the so-called Koch curve (see Fig. 4.5).

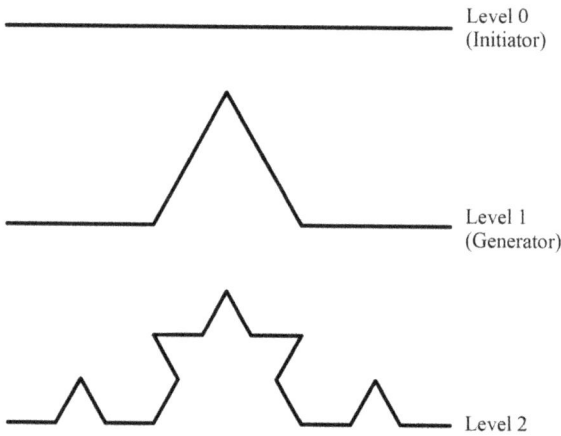

Fig. 4.5: The Koch curve, fractal dimension $D \cong 1.26$.

In determining the fractal dimension of the Koch curve, we can follow the procedure introduced above for setting of the fractal dimension for the Cantor set where rel. (4.8) can be rewritten into the form

$$D = \ln \,(\text{a number of newly created parts}) \,/ \ln \,(1/\text{scaling factor}), \tag{4.9}$$

in the case of the Koch curve

$$D = \ln(4) \,/ \ln(1/(1/3)) = \ln(4) \,/ \ln(3) \cong 1.26. \tag{4.10}$$

Due to self-similarity it does not matter at which level we want to determine a fractal dimension. Equation (4.10) determines the fractal dimension at level 1, and at level 2 we get

$$D = \ln(16) \,/ \ln 1/(1/9)) = \ln\left(4^2\right) \,/ \ln\left(1/\left((1/3)^2\right)\right) = \ln(4) \,/ \ln(3) \tag{4.11}$$

as expected.

The paradox between an area and a perimeter interpreted in the classical (integer) Euclidean geometry can be illustrated on the example of the Sierpiński triangle (gasket, sieve) (see Fig. 4.6). The initiator is represented by a triangle (originally equilateral), and the corresponding generator misses the inner triangle with vertices coinciding with the centres of the initiator edges. Denoting A_n and P_n the area and perimeter of a set at level n, respectively, we obtain

$$A_n = (3/4)^n A_0, P_n = (3/2)^n P_0, \tag{4.12}$$

where A_0 and P_0 represent the area and perimeter of the initiator, respectively. From here it is apparent that for $n \to +\infty$ an area shrinks to 0 while a perimeter tends to infinity. This example substantiates an introduction of the term fractal dimension (in this case, $D = \ln(3)/\ln(1/(1/2)) = \ln(3)/\ln(2) \cong 1.58$) and also documents the same fractal dimension for the topologically similar objects (in this case, any triangle can serve as an initiator with the result $D \cong 1.58$). Usefulness of seemingly abstract constructions of fractal objects can be documented here by gradually lighter objects with an increasing number of branching points contributing to its stiffness (a principle of the Eiffel tour and Eiffel railway bridges).

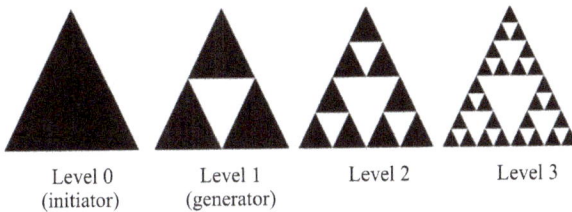

Level 0 (initiator)   Level 1 (generator)   Level 2   Level 3

Fig. 4.6: The Sierpiński triangle, fractal dimension $D \cong 1.58$.

The Menger sponge is an example of an originally 3D object that is transformed to the object with the fractal dimension lower than 3. The initiator is represented by a cube (3 × 3 × 3), and the generator consists of 20 identical sub-cubes that remained after removing the innermost one with its six adjacent neighbours (see Fig. 4.7). Hence, the fractal dimension is $D = \ln(20)/\ln(3) \cong 2.73$ as determined at level 1. Generally, at level n, the fractal dimension $D = \ln(20^n)/\ln(3^n)$ [$=\ln(20)/\ln(3)$, each new cube consists of 20 new sub-cubes].

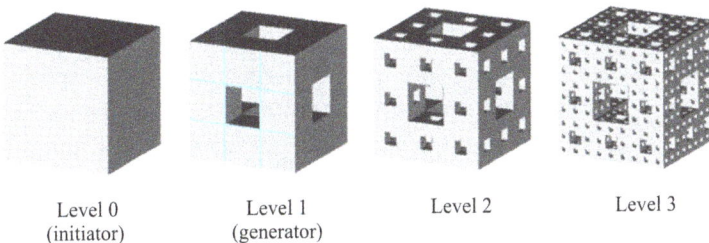

Level 0 (initiator)   Level 1 (generator)   Level 2   Level 3

Fig. 4.7: The Menger sponge, fractal dimension $D \cong 2.73$.

The passages between the individual discrete Euclidean dimensions have not been analysed only in the last 60 years (Mandelbrot 1967, 1982), but the first steps date back to the end of the nineteenth century. The most typical example is the so-called Peano curve (Peano 1890) starting its iterative process with the 1D four linear segments and consecutively filling the 2D square $(0,1) \times (0,1)$ (see Fig. 4.8). In each step the individual (sub)square is divided into nine new ones. The final continuous curve passes through every point of the unit square. More curves are presented in Antoniotti et al. (2020).

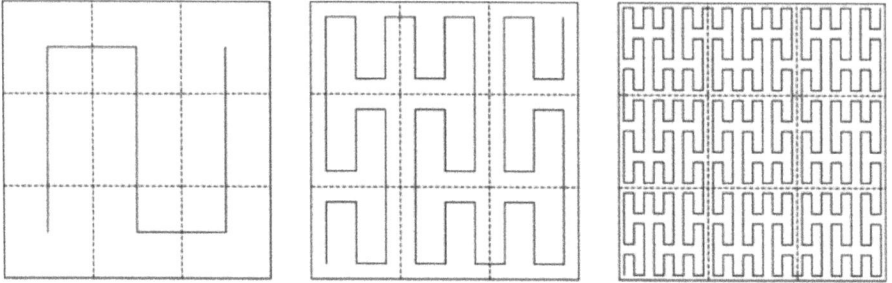

Fig. 4.8: The area-filling Peano curve.

The trajectory of Brownian motion is the example of a curve which 'visits' repeatedly an infinite number of points and simultaneously does not cross an infinite number of points. However, if we choose any point and any neighbourhood of this point, this curve intersects the chosen neighbourhood after a finite number of steps.

The key question is whether the notions of fractal and Euclidean dimensions coincide for the classical objects such as length, area, and volume. If the fractal dimensions for these objects were different, then its introduction would be meaningless. As expected, the answer is affirmative as can be documented on the example of a square (see Fig. 4.9). A number of newly created parts attain 36 and a scaling factor is 1/6. According to eq. (4.9) the fractal dimension

$$D = \ln(36) \ / \ln(1/(1/6)) = \ln\left(6^2\right) \ / \ln(6) = 2 \tag{4.13}$$

corresponds to the Euclidean one. Analogously in 3D space the generator can be represented by 1 small sub-cube out of 27 forming the whole cube (initiator).

The discrete Euclidean and 'continuous' fractal dimensions have one common attribute. If an observer in the Euclidean space (point – perimeter – circle – sphere, point – length – area – volume) is in one of these four possibilities (e.g. at the level of circle), he perceives the 'inferior' objects (point, straight line) as the ones with $D = 0$ and the 'superior' objects (sphere) with $D = \infty$. The exactly same situation is with the fractal dimension. If an observer moves in a fractal dimension D_{obs}, the 'inferior' objects ($D < D_{obs}$) are perceived with $D = 0$, and the 'superior' objects ($D > D_{obs}$) are perceived with $D = +\infty$. It implies that any object exhibits only one finite dimension.

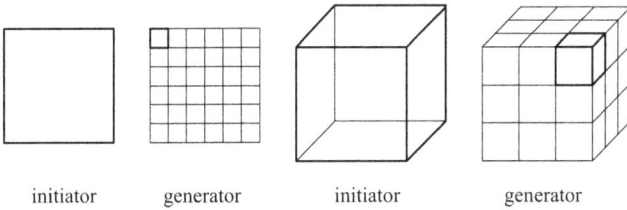

initiator   generator   initiator   generator

Fig. 4.9: Coincidence of fractal and Euclidean dimensions.

Let us very briefly introduce the Hausdorff measure to compare the determination of a fractal dimension through this original approach and the simplified self-similarity approach (rel. (4.4)) above.

If we have any set S, we can define its diameter *diam* as the longest distance ρ of two arbitrary points in S

$$diam(S) = max\{\rho(x,y): x,y \in S\} \tag{4.14}$$

and consequently, to determine for any positive numbers d and δ, the expression

$$\sum_{i=1}^{\infty} (diam\ U_i)^d, \tag{4.15}$$

where a family of subsets U_i ($i = 1, \ldots, \infty$) represent a countable cover of S ($S \subseteq \bigcup_{i=1}^{\infty} U_i$) (for illustration, see Fig. 4.10) and any subset U_i fulfils $diam\ U_i < \delta$. Here $\bigcup_{i=1}^{\infty} U_i$, denoting a union of the collection of subsets U_i is the set of all elements in the collection. In this way we obtain – for any fixed d and δ – a minimum value (or rather infimum as the values asymptotically tend to the lowest value) for any countable covers of S. With a decrease of δ, a number of all possible covers decreases as the cases with at least one subset U_i exceeding $\rho(x,y) > \delta$ are eliminated. This results in an increase of the infimum

$$H_\delta^d(S) = \inf\left\{ \sum_{i=1}^{\infty} (diam\ U_i)^d : S \subseteq \bigcup_{i=1}^{\infty} U_i,\ diamU_i < \delta \right\}. \tag{4.16}$$

For the limiting case $\delta \to 0$ we obtain the Hausdorff outer measure of dimension d of a set S

$$H^d(S) = \lim_{\delta \to 0} H_\delta^d(S). \tag{4.17}$$

The value of d was chosen arbitrary but from the preceding paragraph we know that there is only one d for which a value $H^d(S)$ is finite (for values exceeding d the value $H^d(S)$ is infinite, for lower values the $H^d(S)$ nullifies). This only d is called the Hausdorff dimension.

The above examples are based on a strict self-similarity when an observer has no chance to realise at which level it is located. Nevertheless, the characterisation of vari-

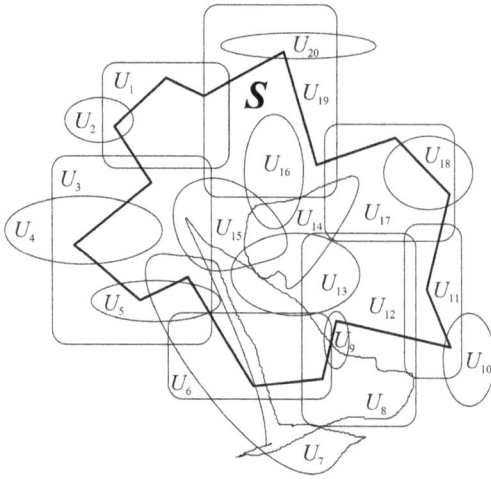

Fig. 4.10: A countable cover of the set S by the family of subsets $\{U_i\}_{i=1}^{\infty}$ (in this case 20 instead of ∞).

ous objects through the fractal dimension proved its applicability in various industrial branches regardless whether they fulfil the mathematical assumptions required for correct introduction of this topic. Usually it concerns the objects that are otherwise very difficult to describe. As the examples it is possible to name, for instance, extruded polymer rods in polymer processing or threads in the textile branch. It is not possible to determine their exact axis and hence their diameter. However, their surface irregularities can be expressed through the notion of fractal dimension. Not surprisingly the fractal characterisation is amply used in connection with flocs. As stated in Jarvis et al. (2008): ". . . optimum floc properties could be determined using the floc fractal dimension combined with the floc size and strength data."

A determination of fractal dimension for flocs cannot follow the procedure outlined above as no self-similarity is attained. The original procedure using the Hausdorff measure is not applicable in practice. Hence, other procedures of determining the fractal dimensions have to be applied as, for instance, a direct calculation using the box counting method or an indirect evaluation based on settling velocity of particles or a method using light scattering as will be presented in the following sections.

4.2 Application of a fractal dimension to coagulation and flocculation

The late 70s and the whole decade of the 80s represent a systematic projection of fractal ideas into various scientific branches such as polymer processing (Tzoganakis et al. 1993; Duan et al. 2021), medicine (Goldberger and West 1987; Ma et al. 2018), geophysics (Meakin 1991; de Arcangelis et al. 2016), and civil engineering (Winslow 1985;

Wang et al. 2021), to name a few. Jointly with the pioneer contributions, the recent reviews are also cited documenting that the fractal applicability is permanently topical.

No wonder that a demonstration of fractal analysis also proved to be useful in the process of coagulation and aggregate-formation (Forrest and Witten 1979; Schaefer et al. 1984; Meakin 1988a). Here it is necessary to remark that the first appearance of a hierarchical model for a variety of particle aggregates with a power (fractal) relation between the mean aggregate density and the aggregate mass ($\rho \sim M^\lambda$) was published already in 1964 (Beeckmans 1964) – 3 years prior to the Mandelbrot's famous work on a fractal classification of the UK coastline (Mandelbrot 1967).

It is not possible to expect that theoretically (purely mathematically) derived fractal formations will have their fractal counterparts in nature. There is a series of reasons preventing coincidence of both categories: purely theoretical one and formations appearing in practice. One of the crucial points is a lower limit of particle size (in the case of colloids or suspensions); in some instances, corresponding to the radius of compact assemblages of smaller particles (Waite 1999), contrasting to infinitesimally thin curves. Even if a process of fractal generation were generated correctly at the first stage, it should be terminated after a few initial steps due to 'bulkiness' of particles.

Another reason is caused by frequent deviation of particles from the spherical shape. A traditional description of colloid (suspension) behaviour supposes a spherical shape ensuring – among other things – that two particles can be simultaneously in touch at one point only. This substantially simplifies modelling of the corresponding behaviour. As the examples two relations can be introduced:
1) The most common model for settling velocity of particles v known as the Stokes' law balancing buoyancy, gravity, and hydrodynamic drag

$$v = \frac{(\rho_p - \rho_c)gd^2}{18\mu},$$
(4.18)

where ρ_p and ρ_c are the mass densities of the particles and the carrier liquid, respectively; g is the gravitational constant, d is the particle diameter (monodispersed spherical particles are supposed), and μ is the dynamic viscosity.
2) The Einstein model (Einstein 1906, 1911) for viscosity of dilute suspensions (concentration ϕ up to 5%) also supposes the spherical shape of particles

$$\mu = \mu_c(1 + 2.5\phi),$$
(4.19)

where μ_c is the viscosity of a carrier liquid.

To apply fractal dimension for a description of colloidal behaviour, a term 'diameter' characterising a shape of spherical particles was generalised for the irregular ones in various ways:
– equivalent sphere diameter of a particle,

– a geometric mean $\sqrt{ab}$ of the major (a) and minor (b) axes of the closest ellipse inscribing the 2D projection of a particle.

For geometrical characterisation of clusters (aggregates) the following terms are widely used:
– a maximum length of a cluster L_{max} representing the longest straight segment within the cluster,
– a perimeter P of a cluster 2D projection,
– very frequently used the radius of gyration expressed as the root-mean-square distance of the clusters forming an aggregate that represents the shortest distance from point mass to the axis of rotation (moment of inertia at that point corresponds to a sum of moments of inertia corresponding to the individual clusters with the masses m_i)

$$R_g = \sqrt{\sum_{i=1}^{n} m_i r_i^2 / \sum_{i=1}^{n} m_i}, \tag{4.20}$$

where r_i represents the shortest distances of the individual clusters from the axis of rotation. In other words, a moment of inertia at the point which distance is R_g from an axis of rotation is identical to the moment of inertia of the whole aggregate (see Fig. 4.11).

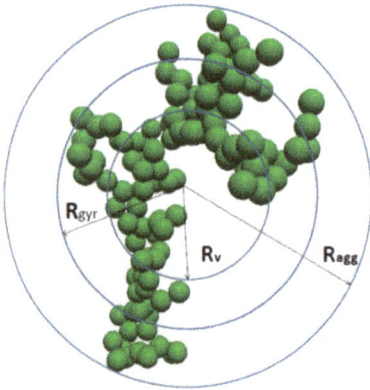

Fig. 4.11: A scheme relating a radius of gyration R_g with a radius of a minimal sphere R_{agg} containing an aggregate and a radius of a minimal sphere with the same material volume as an aggregate (taken from Fard 1984, Fig. 3.1).

A notion of fractal dimension was gradually introduced for characterising of coagulation process through several relations fulfilling the following attribute: a power relation between two physical or geometrical quantities related to the aggregate's description, hence, algebraically similar to rel. (4.3):
– $A \propto r^{D_2}$, where A is the sum of areas of all primary particles contained within a circle of radius r (Chakraborti et al. 2003) and D_2 is the two-dimensional fractal dimension, i.e. $1 < D_2 < 2$;

- $A \propto L_{\max}^{D_2}$, in the case of non-spherical particles, a radius (diameter) is replaced by the longest straight segment $L_{\max}$ (Logan and Kilps 1995; Gorczyca and Ganczarczyk 1996; Chakraborti et al. 2003);
- $P \propto L_{\max}^{D_1}$, D_1 is the one-dimensional fractal dimension, i.e. $0 < D_1 < 1$ (Kilps et al. 1994; Jarvis et al. 2008);
- $M \propto R_g^{D}$, where M is the mass of the fractal aggregates (Waite 1999), this relation implies the following relation:
- $\rho \propto R_g^{D}$, where ρ is the density of aggregates (Waite 1999);
- $N \propto R_g^{D}$, where N is the number of particles per aggregate (Waite 1999);
- $M \propto L_{\max}^{D}$ relating the mass and the characteristic length (Meakin 1988a; Johnson et al. 1996; Bushell et al. 2002), and analogously
- $N \propto L_{\max}^{D}$ under the assumption of comparable individual particles.

As apparent, a variety how to introduce a fractal dimension is relatively wide and the above outline is only illustrative; some relations are sketched in Fig. 4.12. Another fractal relation can be developed, for instance, by involving particle characterisation into the last relations: $N \propto (R_g/r)^{D}$ (Gregory 2009; Lopez-Exposito et al. 2019).

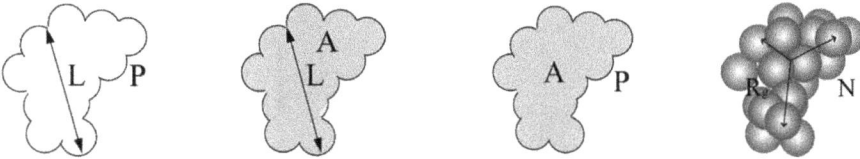

Fig. 4.12: Fractal dimension based on the inter-relations between pairs of geometrical quantities (taken from Lee and Kramer 2004).

An introduction of a term 'fractal dimension' is not only of an academic meaning but has an immediate practical impact. As stated in Jiang and Logan (1991): "In order to describe coagulation by use of fractal geometry, the properties of the fractal aggregates must be cast in terms of fractal dimensions. These properties, such as mass, volume, density, porosity, and settling velocity affect particle collisions and, therefore, coagulation rate."

Non-negligible influence on fractal dimension is how the aggregates are formed. There is a remarkable difference in fractal values if the aggregates are formed by shear coagulation (see Fig. 4.13) or by differential sedimentation. The aggregates exposed to shearing exhibit more compactness (see Fig. 4.14), and hence, higher fractal dimension than those more or less loosely moving (Bubakova et al. 2013). For more compact aggregates, shear forces are high enough to promote aggregate restructuring but not high enough to break primary particles in small fragments (Moruzzi et al. 2017).

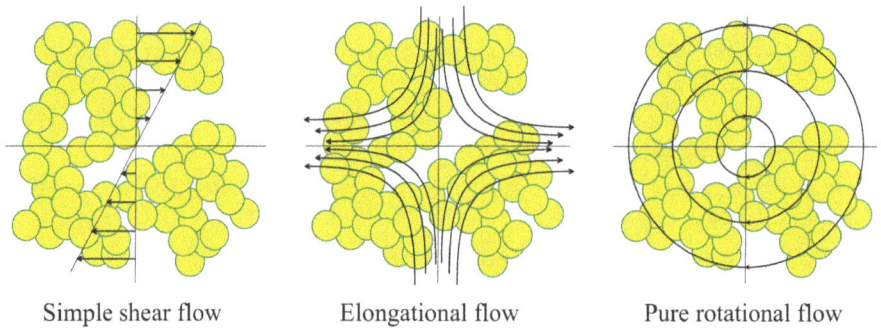

Simple shear flow   Elongational flow   Pure rotational flow

Fig. 4.13: Three types of deformation exerted to an aggregate: simple shear flow, elongational flow, and purely rotational flow (taken from Fard 1984, Fig. 6.15).

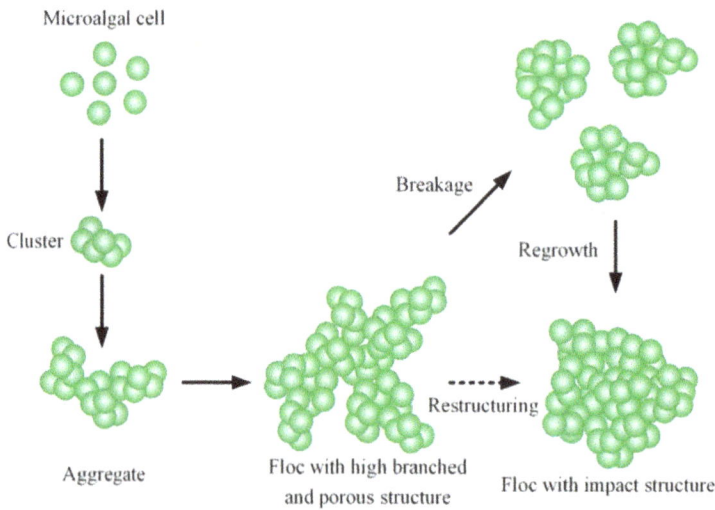

Fig. 4.14: The aggregates undergoing the breakage-regrowth process caused by shearing. This results in a compact structure (taken from Zhang et al. 2019, Fig. 1, where motivated by Fig. 1 in Spicer et al. 1996).

Analogously, in the case of low 1D fractal dimension (e.g. in the case $P \propto L_{\max}^{D_1}$), the lower values of D_1 indicate that the aggregates are more peripheral regular and spherical with smaller specific surface area (Zhao et al. 2013). Figure 4.15 provides a visual comparison for different values of fractal dimension starting from nearly straight chain followed by very loose, open, and stringy structure and terminating in a relatively compact form.

Figure 4.15 clearly documents usefulness of an introduction of the term 'fractal dimension' in characterisation of particles-clusters-aggregates morphology. As expected, practically any of the physical quantities connected with aggregate properties – if expressed through the fractal dimension – provides much better insight (qual-

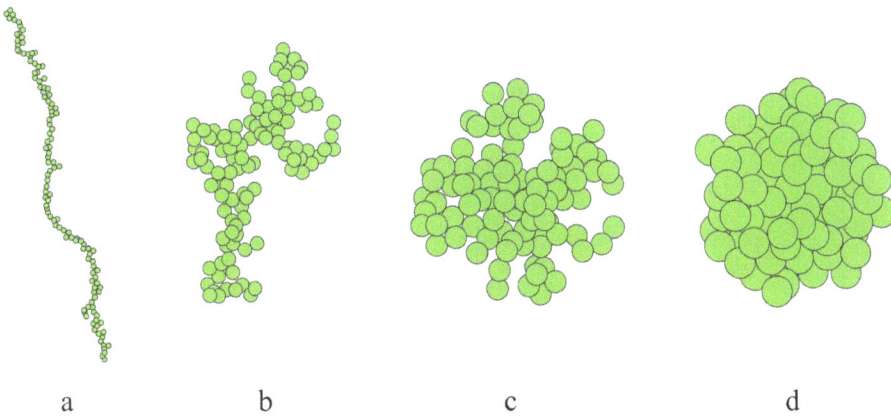

Fig. 4.15: Fractal values for various aggregate forms attaining consecutively: (a) 1.2, (b) 1.85, (c) 2.3, and (d) 2.8 (taken from Fard 1984, Fig. 3.2).

itatively and quantitatively) in an aggregate structure. Let us mention mass, volume, and density. Other quantities, discussed later, will be based on the experimental techniques used for an evaluation of fractal dimension.

A value of fractal dimension is not fixed for the given studied (colloidal, suspended) material but develops in time. First, during the initial perikinetic stage (slow mixing) accompanied by creation of relatively loose clusters a fractal dimension gradually increases. However, in confrontation with higher shearing in the orthokinetic stage, a fractal dimension suddenly steeply drops owing to 'cutting' of loose surface chains (see Fig. 4.14). Finally, relatively compact stabilised aggregate form (see Fig. 4.15) is achieved exhibiting among other things better sedimentation than loose formations. The changing hydromechanical conditions in time are the main impetus (Spicer et al. 1998, Wang et al. 2011a) in a three-stage time-dependent development of aggregate size (see Fig. 4.16: growth – drop – re-growth) and the values of the corresponding fractal dimension. If the flocs are exposed to higher shearing just from their early structuring then a deviation of fractal dimension in time is rather moderate. Otherwise the flocs are susceptible to breakage with increasing shearing which results in apparent fractal changes. The fractal dimension of a floc gives practical information related to porosity, permeability, density, strength, terminal sedimentation velocity, and structure (Kovalsky and Bushell 2005, Lopez-Exposito et al. 2019). For instance, what concerns density; for solid 3D Euclidean objects, a density is completely independent of size. However, for low fractal dimensions there can be a marked decrease in density as aggregates grow larger. Large aggregates can have very low density and this has significant consequences for solid-liquid separation (Gregory 1998). In the fractal framework, a fractal density is also dependent on time.

For monodisperse spherical particles Kranenburg (1994) proposed the following relation:

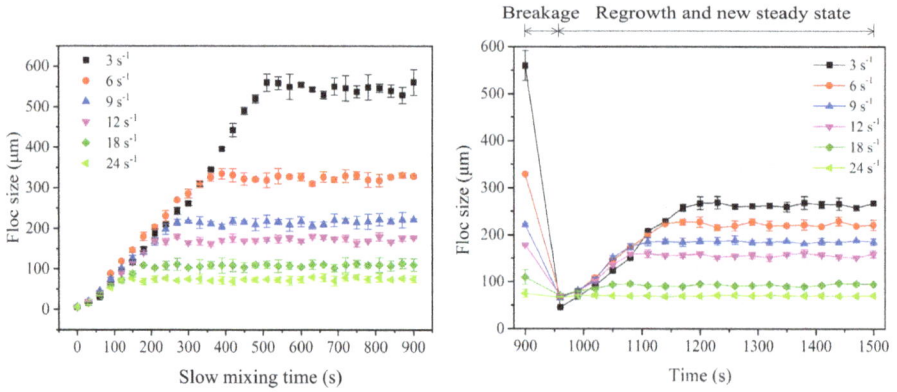

Fig. 4.16: Time-dependent floc size documenting its variance with respect to shear intensity, for higher shear values nearly independent (taken from Zhang et al. 2019; Figs. 2 and 4).

$$\rho_f - \rho_c = (\rho_p - \rho_c)(D/d)^{F-3}, \tag{4.21}$$

where ρ_f, ρ_c, and ρ_p are the densities of floc, carrier liquid, and primary particles, respectively; D is the floc diameter and d is the primary particles diameter. The quantities ρ_c, ρ_p, and d are time-invariant. However, the floc diameter D strongly subjects to time evolution as well as the floc density ρ_f. To overcome the condition of monodisperse spherical particles, Lau and Krishnappan (1994) proposed the following relation with two parameters b and c

$$\rho_f - \rho_c = (\rho_p - \rho_c)\exp(-bD^c), \tag{4.22}$$

Combining these two relations Garcia-Aragon et al. (2014) obtained a relation for the determination of fractal dimension F

$$F = 3 - \left(bD^c / \ln\left(\frac{D}{d}\right)\right), \tag{4.23}$$

in other words, this relation expresses the time evolution of fractal dimension through time evolution of floc sizes. The floc size distribution is influenced by a series of factors including turbulent energy dissipation and concentration (Dyer and Manning 1999) which makes its determination more complicated.

Relation (4.23) demonstrates that a notion 'fractal dimension' is generally introduced and is not interlaced only with an exponent (or its part) in the power-law expressions relating two physical (physico-chemical) quantities. Hence, this term cannot be defined exclusively through a slope in the log-log diagram (as will be discussed below).

Not always a determination of the fractal dimension is an easy step as documented by the Weierstrass-Mandelbrot fractal function (Berry and Lewis 1980)

$$F(x) = \sum_{n=-\infty}^{\infty} b^{-n(2-D_f)} \left[1 - \cos(b^n(x))\right] \tag{4.24}$$

used for modelling of the machine surface profile (Hasegawa et al. 1996). Here, b (>1) is a constant and D_f ($1 < D_f < 2$) is the fractal dimension of the profile. It is not without interest that this continuous function is non-differentiable at all points.

Once the notion of a fractal dimension is adopted, it should not be mixed with the classical Euclidean (discrete) geometry ($D = 0, 1, 2, 3$). For instance, let us start with the expression relating the mass of an aggregate with its radius through the fractal dimension D

$$M \propto R^D \tag{4.25}$$

implying ($M = \rho \times V$)

$$\rho V \propto R^D. \tag{4.26}$$

At this point it is rather speculative to replace V by the Euclidean term $(4/3)\pi R^3$ and to obtain

$$\rho \propto R^{D-3} \tag{4.27}$$

as during this approach (fractal-Euclidean-fractal) there are two passages: one from the space of continual (fractal) dimension to the space of discrete (Euclidean) dimensions and the other one going in the opposite direction.

The term 'fractal dimension' is naturally interlaced with other quantities characterising aggregate structure as, for instance, strength. The inter-particle bonds between the components of the aggregate govern floc strength (Bache et al. 1997; Boller and Blaser 1998). The resistance of a floc against breakage is called the floc strength and is evaluated by the ratio (see Fig. 4.15):

$$\text{Strength factor} = \frac{d_{in}}{d_{br}} \times 100\,\%, \tag{4.28}$$

where d_{in} and d_{br} denote the sizes of initial and broken flocs, respectively (Francois 1987). The re-aggregating capacity of the broken flocs is expressed through the so-called recovery factor

$$\text{Recovery factor} = \frac{d_{ra} - d_{br}}{d_{in} - d_{br}} \times 100\,\%, \tag{4.29}$$

where d_{ra} is the re-aggregated floc size.

As apparent from the preceding two relations a time course of fractal dimension roughly corresponds to a course of a strength factor. There are two modes of floc rupture (Parker et al. 1972): (1) surface erosion removing small particles from the floc surface corresponding to tangential shearing and (2) large-scale fragmentation when tensile strength acting normally causes disintegration of a floc. The first case occurs in the case of a compact floc (higher strength and higher fractal dimension), and the second one corresponds to the looser floc structure (lower strength and lower fractal dimension).

From above we can derive that from a time course of floc sizes (characterised among other things also by the strength and recovery factors) it is possible to obtain a rough estimate of a time course of fractal dimensions and vice versa. Fractal dimension, floc size, and strength data dominantly participate in floc characterisation.

4.3 Evaluation of a fractal dimension

An overwhelming majority of expressions defining a fractal dimension relate to two physical quantities through power-law dependence as documented above. If we denote these quantities as Q_1 and Q_2, then the power-law relation is of the form

$$Q_1 = C \times Q_2^{D_f}, \tag{4.30}$$

where D_f is the fractal dimension and C is the multiplicative constant. By far the seemingly simplest way how to determine a fractal dimension is to logarithmise rel. (4.30)

$$\log(Q_1) = \log(C) + D_f \times \log(Q_2). \tag{4.31}$$

The multiplicative term C converts in the log-log coordinates into an additive term log (C), and D_f in this coordinate system represents a slope of a linear straight line Q_1 vs. Q_2. If we have a set of the experimental data $\{[Q_1^i, Q_2^i]\}_{i=1}^n$, then n experimental couples are transformed to a set $\{[\log(Q_1^i), \log(Q_2^i)]\}_{i=1}^n$. Nevertheless, this transformation is rather speculative as the additive term log(C) optimised in the log-log coordinates has no physical meaning, and seemingly correct (from the visual viewpoint) approximation of D_f in the log-log coordinates can be questioned in the normal coordinates. In other words, the linear regression in the log-log coordinates

$$D_f = \frac{\{n \sum [\log(Q_1^i) \times \log(Q_2^i)]\} - \{\sum \log(Q_1^i) \times \sum \log(Q_2^i)\}}{n \sum (\log(Q_2^i))^2 - (\sum (\log(Q_2^i)))^2}, \tag{4.32}$$

$$\log(C) = \frac{\{\sum \log(Q_1^i) \times \sum (\log(Q_2^i))^2\} - \{\sum \log(Q_2^i) \times \sum [\log(Q_1^i) \times \log(Q_2^i)]\}}{n \sum (\log(Q_2^i))^2 - (\sum \log(Q_2^i))^2} \tag{4.33}$$

minimises the mean deviation of the experimental data from an optimised linear straight line (a summation is implicitly supposed from 1 to n). In contrast to it, a nonlinear regression in the original normal coordinate system optimises the parameters D_f and C in such a way that a mean deviation of the experimental couples $\{[Q_1^i, Q_2^i]\}_{i=1}^n$ is calculated with respect to a supposed power-law (not linear) function.

In this case the parameters D_f and C can be calculated based on the following procedure.

Nonlinear regression optimising simultaneously both parameters uses the method of least squares which minimises the sum of the squares of the residuals (the differences between the experimental data and the predicted power-law model values). To this aim the data are fitted by a method of successive approximations. Each successive approximation is based on the Newton's (Newton-Raphson) method, which principle is shown in Fig. 4.17.

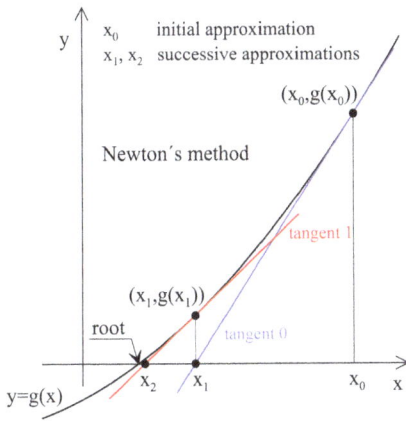

Fig. 4.17: A sketch of successive approximations providing an approximate root of the equation $g(x) = 0$.

If we want to determine a root of the function $g(x)$, we choose the initial approximation of the root x_{root}; let us say x_0, and the first approximation x_1 is calculated from the relation

$$x_1 = x_0 - \frac{g(x_0)}{g'(x_0)} \tag{4.34}$$

as an intersection of a tangent of $g(x)$ at the point $(x_0, g(x_0))$ and the axis x. This way it is possible to proceed with x_2, x_3, etc. gradually approaching a root of the function $g(x)$ with the prescribed deviation. Hence, it is necessary to determine a derivative of the function $g(x)$ at all consecutive points x_n

$$x_{n+1} = x_n - \frac{g(x_n)}{g'(x_n)}. \tag{4.35}$$

Sometimes it is useful – for a general course of $g(x)$ – to shorten a shift $(x_{n+1} - x_n)$ in rel. (4.35) not to jump over the root.

For the case of a power-law function $y = Cx^D$ (analogous to rel. (4.30)) we want to minimise a sum of squares of the differences between the experimental data y_i and the predicted values Cx_i^D

$$S(C, D) = \sum_{i=1}^{n} (Cx_i^D - y_i)^2. \tag{4.36}$$

As the sum should attain a minimum value, the following two conditions have to be fulfilled

$$\frac{\partial S(C, D)}{\partial C} = 0, \qquad \frac{\partial S(C, D)}{\partial D} = 0. \tag{4.37a, b}$$

From the first condition we obtain

$$C(D) = \frac{\sum\limits_{i=1}^{n} y_i x_i^D}{\sum\limits_{i=1}^{n} x_i^{2D}}. \tag{4.38}$$

The second condition implies (where $F(D)$ denotes a derivative of S with respect to D)

$$F(D) \equiv \sum_{i=1}^{n} \ln(x_i) \times \left[C^2 x_i^{2D} - Cy_i x_i^D \right] = 0. \tag{4.39}$$

The function $F(D)$ takes over a position of the function $g(x)$ above. To determine successive points D_n converging to an optimal value D, it is necessary to determine a derivative

$$F'(D) = \sum_{i=1}^{n} \ln(x_i) \times \left[2Cx_i^{2D} \frac{dC(D)}{dD} + 2C^2 x_i^{2D} \ln(x_i) - \frac{dC(D)}{dD} y_i x_i^D - Cy_i x_i^D \ln(x_i) \right], \tag{4.40}$$

where a derivative of $C(D)$ with respect to a fractal dimension D follows from rel. (4.38)

$$\frac{dC(D)}{dD} = \frac{\left[\left(\sum\limits_{i=1}^{n} y_i x_i^D \ln(x_i) \right) \times \left(\sum\limits_{i=1}^{n} x_i^{2D} \right) \right] - 2 \left[\left(\sum\limits_{i=1}^{n} y_i x_i^D \right) \times \left(\sum\limits_{i=1}^{n} x_i^{2D} \ln(x_i) \right) \right]}{\left(\sum\limits_{i=1}^{n} x_i^{2D} \right)^2}. \tag{4.41}$$

To summarise, the whole procedure starts with a choice of the initial value D_0 and its insertion into the rels. (4.51)–(4.54). This initial entry input can be based on the usage of rels. (4.32) and (4.33) corresponding to the linearised regression. Then the first approximation D_1 can be calculated from an analogue to rel. (4.35), where a function g is substituted by $F(D)$. The obtained D_1 is again substituted into rels. (4.38)–(4.41) to calculate D_2, etc. The whole process can be stopped when a deviation of two successive approximations is not significant. It is necessary to take into consideration that the deviations of two successive D's should be much lower than those for C's as a power function is relatively sensitive to the changes in its exponent. It is also apparent that the whole procedure can be very simply solved in any spreadsheet as the dominant portion – data input – is the same as in the case when rel. (4.30) is logarithmised to rel. (4.31).

The nonlinear regression (and other computational techniques processing the power-law relation) is in a compliance with a definition of fractal dimension how it is introduced in the process of coagulation and should be used for the determination of fractal dimension. The differences between the two procedures, an optimisation in the log-log coordinates (with the parameters $\log(C_{lin})$ and D_{lin}) and an optimisation in the normal (Cartesian) coordinates (with the parameters C_{lin} and D_{lin}), are illustrated in Fig. 4.18 relating an area A with a characteristic length L ($A = C \times L^D$). The black straight line corresponds to a determination of (C,D) using an optimisation in the ln-ln coordinates for which $\ln(C_{lin}) = -1.214$ ($C_{lin} = 0.297$) and $D_{lin} \cong 1.603$ and the blue line to the determination through an optimisation in the physically correct normal coordinates for which $C_{nonlin} = 0.190$ ($\ln(C_{nonlin}) = -1.661$) and $D_{nonlin} \cong 1.652$. It seems visually that the blue line has no connection with the experimental data. However, if we compare the mean deviations for the couples ($C_{nonlin} = 0.190$, $D_{nonlin} \cong 1.652$) and ($C_{lin} = 0.297$, $D_{lin} \cong 1.603$), then we obtain two rather different mean deviations of the individual experimental points from the definitory curve $A = C \times L^D$ for the fractal dimension D: for the couple corresponding to the optimisation in the normal coordinates 40.6% and for the couple obtained in optimising the ln-ln coordinates 43%.

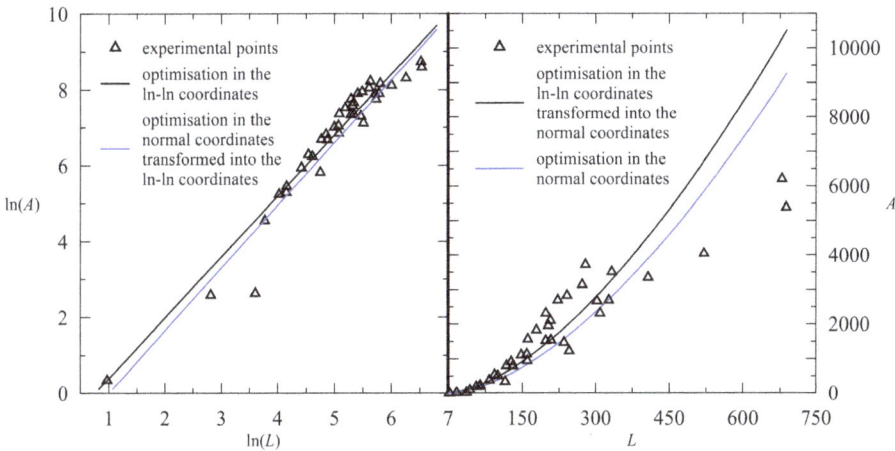

Fig. 4.18: A comparison of the courses of linearised (black) and true (blue) approximations of a fractal dimension (data for Area C taken from Zheng et al. 2011, Fig. 9).

It is necessary to realise that the differences in fractal dimensions cannot be compared with the whole interval (0,2) but only with respect to the expected interval; in this case approximately (1.5,1.95). Hence, the difference (~0.05) is not negligible within the range 1.5 ÷ 1.95. This can also be illustrated by a comparison of the round Szierpiń-ski carpets depicted in Fig. 4.19. The difference in fractal dimensions of both objects attains 0.08 (left object 1.626 and right object 1.713). If the values of both optimised fractal dimensions are inserted into a sequence 1.603 – 1.626 – 1.652 – 1.713, then we can see that a difference (also visual) between the fractal dimensions 1.603 and 1.652 is apparent.

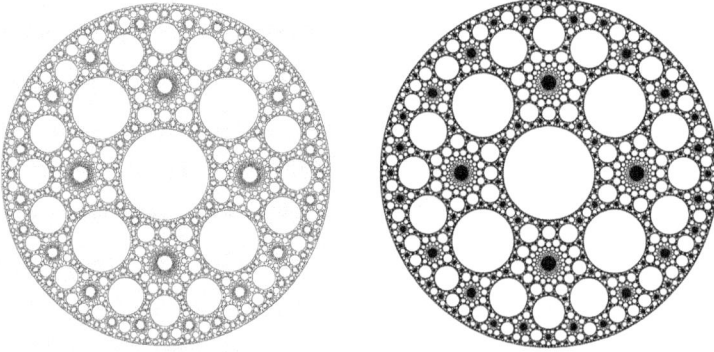

Fig. 4.19: The modified versions of the round Szierpiński carpet, the fractal dimensions of 'empty' (left) and 'full' (right) carpets ≅ 1.626 and ≅ 1.713, respectively (taken from Moghaddam and Cheriet 2015, Fig. 2).

Discrepancy between the logarithmic and normal optimisations is more evident in the following example depicted in Fig. 4.20. If – in the log-log coordinates – the data is uniformly deviated from a straight line, it optically evokes an impression of equidistant differences of the individual points from the straight line. In fact, for such data the percentage deviation is constant and invariant with respect to the log(x) value as follows from the following calculation:

$$100 \times \left(\frac{y_1 - y_2}{y_1}\right) = 100 \times \left(\frac{Cx^D - \left(1 - \frac{\text{dev}}{100}\right)Cx^D}{Cx^D}\right) = \text{dev.} \qquad (4.42)$$

The situation differs for the normal coordinates. A constant percentage deviation is accompanied with still higher distances from the steeply increasing basic curve for the increasing x-values contrasting to the dramatic decrease in deviations of a constant additive increment with an increasing x-location. This is the principal issue why visual approaches to the log-log and normal coordinates have to be taken in different ways.

Better insight into the differences in deviations transformed by a passage from the normal (linear) coordinates into the log-log coordinates is depicted in Fig. 4.21. This documents different optimisation in both coordinates.

The comparison of both optimisations differing in the applied coordinates provides the following conclusions:

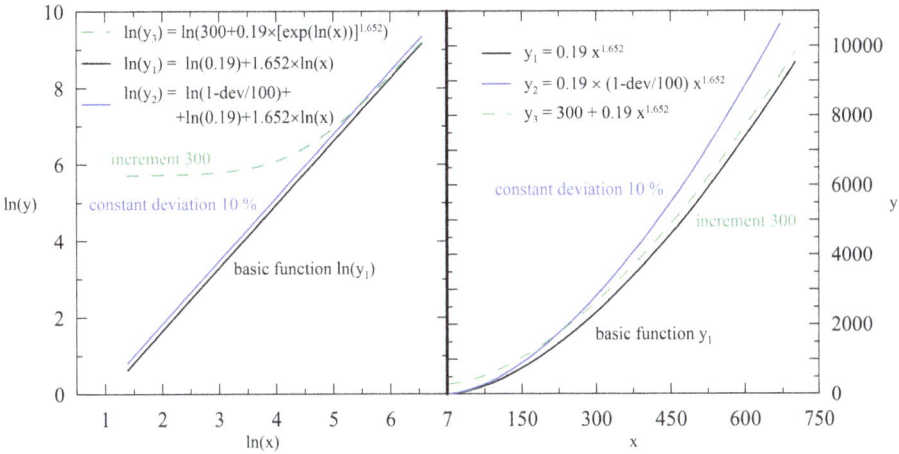

Fig. 4.20: The curves y_2 represent a constant percentage deviation (in this case 10%) from the basic curve y_1, in the log-log coordinates the curves $\ln(y_1)$ and $\ln(y_2)$ are always parallel. The parallel curves in the normal coordinates (y_1 and y_3) correspond to a constant increment (in this case 300).

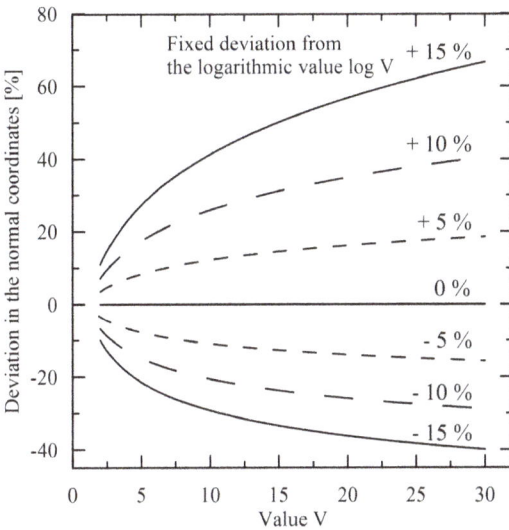

Fig. 4.21: A comparison of the deviations in the normal and log-log coordinates.

- relatively close linearised fractal dimensions have not to be so close in true fractal dimensions, which can influence potential conclusions;
- traditionally accepted intervals for flocs with various degrees of compactness can be shifted;
- any physical interpretation of the constant $\ln(C)$ is of a limited value.

4.4 Determination of a fractal dimension

There exist a series of experimental techniques providing various characteristics of coagulation processes. The problem is that practically all of them are 'balanced' what concerns their 'pros' and 'cons'. This is – among other things – influenced by these three crucial points:

1) A range of particle sizes is very diverse (a basic distribution starts with colloidal non-aggregated particles (up to ≈5 μm), larger primary- and micro-aggregates (up to ≈1 mm), and finally macro-aggregates (exceeding 1 mm). In other words, a ratio of colloidal and macro-aggregate sizes is lower than 0.001.
2) Flocs do not represent static objects but just reversely, their characterisation is strongly time-dependent including values of a fractal dimension as documented above. Capturing the characteristics at one instant can cause difficulties.
3) The aggregates are not compact solid objects but represent relatively flexible objects.

These factors strongly participate in a choice whether it is possible to apply intrusive or non-intrusive techniques for aggregate characterisation. A selection of these methods is presented in the following sections.

4.4.1 Initial fractal approaches

Probably the first work contributing to the introduction of a fractal dimension dates back already to 1964 (Beeckmans 1964). However, it took another 15 years when appeared the first experimental work intentionally devoted to the measurements of material characteristics and their fractal interpretation (Forrest and Witten 1979). The 80s already represented a great boom of this topic on the experimental and simultaneously on the computer scenes. Two interlaced methods on a determination of fractal dimension exclusively used at the beginning of this period are presented below.

4.4.1.1 Box counting method
We will concentrate to the objects with fractal dimensions in the interval [1,2]. The box counting method in its simplest way is based on a discretisation of the studied object; in other words, we cover the object by the regular square grid with a uniform length size δ of each mesh. Let us denote $N_\delta(S)$ a number of squares having non-empty intersection with the object S. Using an analogue to rel. (4.3)

$$N_\delta(S) = \text{Const} \times \delta^{-D_\delta(S)}, \tag{4.43}$$

we obtain

$$\ln\left(N_\delta(S) = \text{Const} \times \delta^{-D_\delta(S)}, (S)\right) = \ln(\text{Const}) - D_\delta(S) \times \ln(\delta). \tag{4.44}$$

Then the box-counting dimension $D_B(S)$ is defined as

$$D_B(S) = -\lim_{\delta \to 0} \frac{\ln(N_\delta(S))}{\ln(\delta)}. \tag{4.45}$$

In practice, it is unrealisable to determine a limiting value for $\delta \to 0$. The simplest way to evaluate $D_B(S)$ is calculating $N_{\delta_i}(S)$ for a couple of the lengths δ_i ($i = 1, 2, \ldots, n$) and to evaluate $D_B(S)$ using rel. (4.44) as a slope of a linear regression curve (a linear straight line governed by the least square method) approximating the computed couples $\left(N_{\delta_i}(S), \delta_i\right)$. This procedure is shown in Fig. 4.22, where a wavy curve represents the studied fractal object. The box-counting fractal dimension $D_B(S)$ is estimated by the relation

$$D_B(S) \cong \frac{\sum\limits_{i=1}^{n} \ln(\delta_i) \times \sum\limits_{i=1}^{n} \ln(N_{\delta_i}(S)) - n \sum\limits_{i=1}^{n} \ln(\delta_i) \times \ln(N_{\delta_i}(S))}{n \sum\limits_{i=1}^{n} (\ln(\delta_i))^2 - \left(\sum\limits_{i=1}^{n} \ln(\delta_i)\right)^2} \tag{4.46}$$

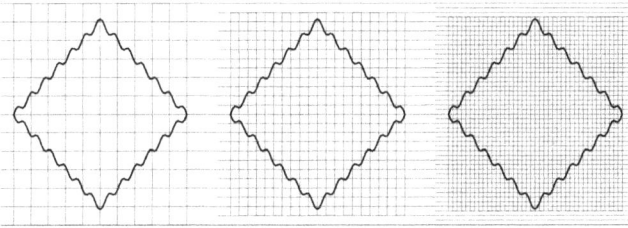

Fig. 4.22: A curve S with the grids of mesh lengths 1, 1/2, and 1/4.

On the same basis as used in Fig. 4.9 it is apparent (and inevitable) that the box-counting dimension coincides with the Euclidean one for $D_B(S)$ equal to 1 and 2. For a point with the Euclidean dimension 0 the numerator in eq. (4.46) nullifies as $N_{\delta_i}(S)$ is identical for all i, and hence the box-counting dimension is 0 as well.

In applying the box-counting method, the used 'covers' can be represented by various forms of a basic element as documented in Fig. 4.23 (Falconer 1990).

Figure 4.24 documents an evolution of (square) pixel coverage of an object (a computer-generated floc – diffusion-limited colloid aggregation) (Bushell et al. 2002). The corresponding fractal dimension determined after six steps (resolutions in pixels vs. number of covering squares: (32, 131), (16, 386), (8, 1198), (4, 3816), (2, 12578), and (1, 42397)) attains 1.67.

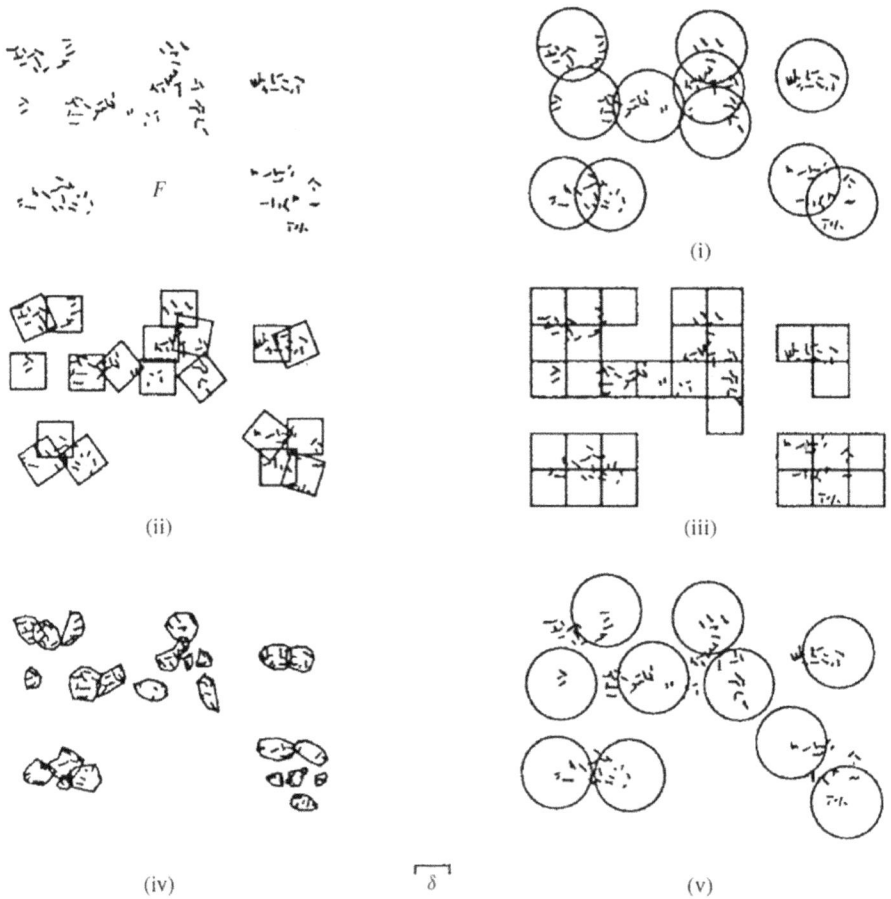

Fig. 4.23: The equivalent 'covers' for a determination of box-counting dimension: (i) the smallest number of closed spheres (circles) of radius δ that cover S; (ii) the smallest number of cubes (squares) of side δ that cover S; (iii) the number of δ-mesh cubes (squares) that intersect S; (iv) the smallest number of sets of diameters at most δ that cover S; (v) the largest number of disjoint spheres (squares) of radius δ with centres in S (taken from Falconer 1990, Fig. 3.2).

It is necessary to emphasise that a box-counting dimension usually does not coincide with the fractal (Hausdorff) dimension (for non-integers) owing to an application to the non-fractal character of the studied objects (as mentioned above) and a significant simplification of the evaluation (4.45) with the help of rel. (4.46). There are also a series of alternatives and improvements to the basic box-counting methods aiming to precise an obtained dimension (Buczkowski et al. 1998; Li et al. 2009; Liu et al. 2014; Nayak et al. 2019; Panigrahy et al. 2019; Wu et al. 2020, 2021; and references therein). An introduction of the box-counting procedure has proven to be helpful in practice for a basic orientation as an appropriate method of fractal dimension estimation for images with or without self-similarity.

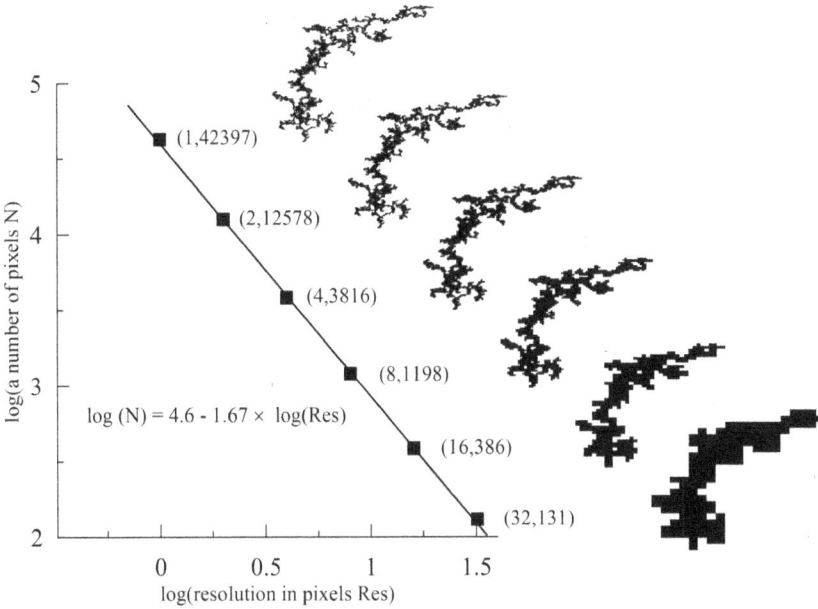

Fig. 4.24: An application of the box counting method to a computer-generated floc (pixel pictures taken from Bushell et al. 2002).

On the other hand, it is necessary to be careful with drawing the conclusions as documented by Fig. 4.25 emphasising the dependence of $D_B(S)$ on a floc 2D-projection. The digital image does not contain only pixels belonging to a one specific cross-section but is composed of all pixels projected in a chosen direction. This means that a projected area is practically always enlarged and this is accompanied – hand in hand – by changes in morphology. Hence, any values of fractal dimension exceeding a value of 2 are false, and the following relation has to be taken into mind (Tence et al. 1986; Thill et al. 1998)

$$D_B = \min\{D_B,\ 2\}. \tag{4.47}$$

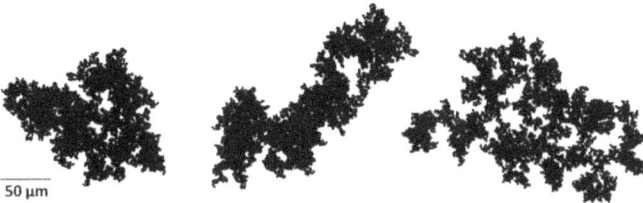

Fig. 4.25: Three projections of the same floc after applying random rotations (taken from Lopez-Exposito et al. 2019, Fig. 6).

4.4.1.2 Sand box technique – nesting square method

Already, in 1979, Forrest and Witten (1979) were probably the first who explicitly carried out the experiments with the aim to calculate a fractal dimension. They worked with micrographs of smoke-particle aggregates and showed an existence of long-range correlations between particles density and their position obeying the power-law dependence. The power-law exponent characterising specific aggregation mechanism represents a fractal dimension. They determined this exponent using the so-called sand box technique (see Fig. 4.26). They consequently counted the symbols in the individual changing squares and obtained a linear relation in the log-log coordinates determining a power-law exponent. This method was later developed by Tence et al. (1986) who used the term 'nesting square method'.

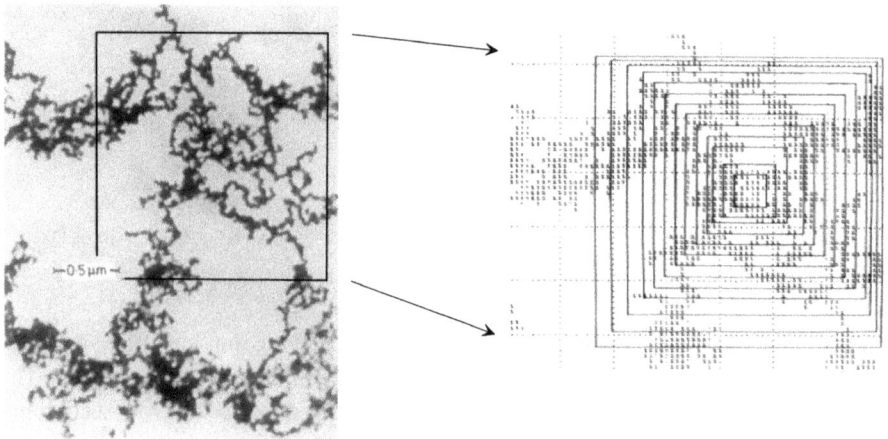

Fig. 4.26: An application of the sand box technique (taken from Forrest and Witten 1979, Fig. 1).

4.4.2 Electrical and optical methods

With the development of experimental technique (electron microscopy, laser technology, electronic devices) a possibility to acquire more sophisticated and accurate experimental data has substantially improved, and more ingenious approaches to data acquisition are at disposal. The improved attributes have solved various aspects of experimental peculiarities but some problems generated by three points summarised at the beginning of Section 4.4 still persist. Moreover, the aggregates are relatively fragile. This implies a necessity to draw attention to non-invasive methods. This condition has two aspects. First, an experimental apparatus should be located outside suspension flow. Second, a geometrical arrangement of flow conditions should not expose the aggregates to any additional shearing. As we will see, not always this condition is fulfilled. Every experimental method exhibits some shortcomings and the individual methods differ in their range. This does not imply that some methods should be a priori eliminated but it is necessary to realise where to pay attention and analyse the results more carefully.

4.4.2.1 Electrical methods

Electrical sensing zone technique (abbreviated as electrozone sensing technique) was developed by Coulter (1953) originally for counting the number of erythrocytes in blood. The so-called Coulter counter (its development in time described in Graham 2003) composes of two chambers separated by narrow cylindrical microchannels (in μm, maximally in units of millimetres). A conductive liquid (electrolyte) is used as a carrier liquid for the particles to be summed up. Two electrodes (one per a chamber) ensure a constant electrical current. Each particle during its passage through a microchannel displaces a liquid volume corresponding to its real volume which reflects in a change of electrical resistance or impedance between both electrodes. In this way not only a passage of particles through a microchannel can be detected but also their sizes differing with respect to intensity of the voltage pulses. A principle of the Coulter counter is shown in Fig. 4.27.

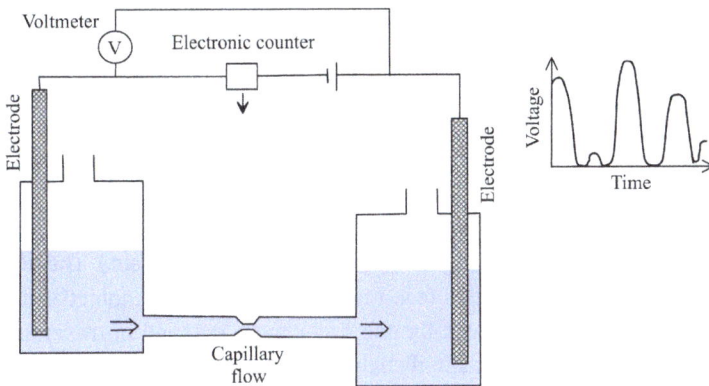

Fig. 4.27: A sketch of the Coulter counter.

The technique is invariant with respect to optical properties (e.g. refractive index) and morphology (shape) of the particles. The great advantage is in evaluation of the real volume of the particles (convenient for fractal analysis) which concentration should be kept in a relatively dilute level. This has two reasons. First, low concentration prevents the particles from blockage of the microchannels. On the other hand, it can contribute to aggregate instability as the particles are exposed to shearing during their passage through the microchannels. Higher counting rate is accompanied by a higher shearing. Secondly, low concentration also improves an accuracy of particles counting as probability of simultaneous 'tandem' passage of two particles is lowered.

A significant shortcoming from the position of drinking water treatment is a lower size limit of particles for which this method can be applied.

The electrozone sensing technique is a very good instrument for counting particles which sizes exceed approximately 0.5 μm but not applicable for lower particles. There is a relation between a used microchannel diameter d and the corresponding sizes of particles that should be within the interval $(0.02 \div 0.6) \times d$. An aperture size of the microchannels usually starts from 20 μm. It means that according to a classical distribution of particles into four groups (non-aggregated colloid particles, primary particles, micro- and macro-aggregates), a group with colloidal particles (from 'zero' to at most 5 μm) cannot be fully detected.

4.4.2.2 Optical methods used in fractal analysis

The onset of relatively new optical methods (scanning and transmission electron microscopy, laser technology) in combination with developing computer industry ('no restrictions' of amounts of data) and electronics has permanently enabled a development of sophisticated and accurate measurements of the characteristics of colloidal suspensions. A crucial aspect in evaluating coagulation characteristics is a degree of preserving hydrodynamic conditions of the studied material. This can be strongly influenced by the two pairs of experimental approaches:

1) Intrusive measurements in contrast to the non-intrusive ones can significantly influence the hydrodynamic conditions and thus non-negligibly affect a process of aggregate creation growth-breakage-regrowth. It is necessary to analyse to which extent such measurements can modify the real conditions.

2) A process of sampling and consequent sample analysis can dramatically change material description (changes caused by possible shearing, dilution, etc.). The advantages acquired by these procedures (e.g. resolution – units of nanometres in scanning electron microscopes, or even by one order more detailed in transmission electron microscopes) can be easily disbalanced by non-corresponding flow conditions and aggregate distortion or breakage.

The following sections will present a selection of the individual techniques related to the measurements of the individual characteristics with an emphasis to their usage in fractal determination.

4.4.2.2.1 Digital image analysis

Digital image analysis was the first analysis applied in the initial fractal approaches (box counting method and sand box technique). Digitising is either direct or initially an analogue picture divided into a (usually) rectangular grid formed by basic squares. An analogue-to-digital convertor assigns to each square – a pixel – information about the brightness of a given colour and opacity. Sharpness of edges in a digitised picture subjects – among other things – to a number of pixels in the horizontal and vertical directions. This is given by the used semiconductor image sensors incorporated in the image capture devices, often used charge-coupled device and now dominating complementary metal-oxide-semiconductor. Storage space occupied by one picture is given by a product of image resolution (a number of pixels, e.g. $8,688 \times 5,792$) and pixel information.

Basically, here are three scales differing in quality of pixel information. A greyscale assigns to a pixel a value within an interval [0,255], where 0 is completely black, 255 is completely white, and the values in between represent the shades of grey. RGB scale information occupies much more space as colour information is based on three integers between 0 and 255 (the integers represent the intensity of red, green, and blue). Hence, a black pixel corresponds to (0,0,0), red to (255,0,0), green to (0,255,0), blue to (0,0,255), and white to (255,255,255). Any combination of numbers in between gives rise to all the different colours existing in nature. The most informative is the RGBA scale extending the RGB scale with an added alpha field representing the opacity of the image (0–100%). On the other side there is a binary (0–1) format, where only the limiting values of the greyscale are included (0 – black, 1 – white). This format can be applied only locally.

Due to various factors (such as quality of illumination, optical quality of lenses, and motion of scanned objects) it is necessary to process the individual pictures (digital description per pixel). Digital processing (in other words software packages) cannot be universally objective but every software routines are governed by subjective decisions (in removing noise, sharpening edges, and counting of objects).

From the viewpoint of an evaluation of fractal dimension, a number of pixels both in a horizontal and in a vertical direction generate a maximal number of steps in the box counting method (see Fig. 4.24). If the basic squares represent (in a reverse order) 1×1, 2×2, 4×4, 8×8, 16×16, 32×32, 64×64, 128×128 pixels, then a relatively dense image $8,688 \times 5,792$ pixels can be in the first approximation covered by $\approx 67 \times 45$ non-covering squares (128×128). However, prior to applying the box counting method it is necessary to clean and rectify a digitised picture to its more realistic (error-free) form.

Generally, digital image analysis (software application) proceeds roughly in four consecutive steps (Merkus 2009). Each of them – as already stated above – is influenced by a subjective role of an experimenter and used software package.

First step – image acquisition – is the most important one. Whatever is spoilt in this step cannot be improved in the consequent ones. A nature of the experiment should dictate a choice of an image capture device and its fixing or relative motion with respect to a course of experiment, remote shutter control, shutter speed, depth of field (the area of apparent sharpness in an image), a selection of the colour scale (quite often used and sufficient 8-bit greyscale), real dimensions of a picture, contrast of a background, illumination, and frequency of pictures. Figure 4.28 was taken in the RAW format, i.e. information about each pixel was deposited individually, and no pixels were joined (accumulated) as is characteristic, for instance, for the JPG format.

Other formats than the RAW one requisition less space (smaller file sizes) but only owing to the lossy compression algorithms (as the discrete cosine transform in the case of the JPG format), which irreversibly lose information about the individual separated pixels.

Fig. 4.28: RAW picture (file) of aggregates in suspension formed by AOM cyanobacterium *Microcystis aeruginosa* and iron coagulant taken in the Taylor-Couette reactor at $G = 20$ s^{-1}.

Second step – image restoration – is oriented to rectification of imperfections accompanying picture capturing in the first step. It is possible to name optical distortion, motion blur, non-uniform illumination, denoising (probably the most subjective element), and gamma correction (raising each pixel value to a certain power (gamma), which helps control the overall brightness and contrast, particularly in low-light or high-light regions), brightness (enhancing or dimming specific areas of an image), and contrast (redistribution of pixel intensities, stretching the intensity range making dark areas darker and bright areas brighter). These corrections also improve edge detection and hence, more accurate determination of box-counting dimension. During this pre-processing step the RAW format has to be still preserved (see Fig. 4.29).

Third step – segmentation and filtering – aims at distinguishing among various specific objects. This is usually achieved by the so-called (binary) thresholding. This technique (in the case of a greyscale) chooses a value between 0 and 255 (optimally based on a histogram of grey values) and assigns a value 0 (black) to pixels which shades of grey are below the threshold level and 255 (white) to the others (exceeding

Fig. 4.29: A pre-processed picture (file) from Fig. 4.28.

the threshold value) (see Fig. 4.30). In the case of touching particles additional techniques can be supplemented (dilation, erosion) to open up the interstitial distances. Here, the binary scale has its justification. As the images are not taken uniformly across their whole area, it can be beneficial to segment the whole area into the individual segments according to their characterisation and to diversify the threshold levels.

Finally, the fourth step following segmentation and filtering makes it possible to characterise quantitatively the studied suspensions. This represents counting of particles and aggregates, determining their morphology, measurement of particles perimeters and areas, and other geometrical quantities (see Fig. 4.31). There exist a series of commercial software packages analysing the suspension characteristics; however, personal supervision is recommended as still usually 3D-objects are projected into only 2D-objects.

The process of image analysis converts 3-dimensionality of the studied objects (flocs, aggregates) to their 2D projections regardless of porosity and position of the individual cross-sections. Further, the individual objects cannot be always mutually distinguished and may be considered as the only one. This point is closely connected with filtering of noise as some pixels regarded as superfluous belong to the object hidden behind the studied one, and on the contrary, some pixels taken as an integral part of the studied object can actually belong to the hidden one. Three-dimensionality and multi-layer arrangement make image analysis more controversial; in some cases,

Fig. 4.30: Image segmentation: thresholding applied to the pre-processed Fig. 4.29.

the classical application (as fractal dimension of the islands) comes out from 2-dimensionality and single-layer topology. All these shortcomings motivated an introduction of the so-called confocal scanning laser microscopy (CSLM) to a description of aggregate development as presented in the next sub-section.

4.4.2.2.2 Confocal scanning laser microscopy

The devices for the measurements based on the confocal microscopy compose of pinholes, laser illumination, objective lenses, detectors (usually low-noise photomultiplier tubes converting light energy into electrical energy), fast scanning mirrors, and filters for wavelength selection (see Fig. 4.32). In contrast to the light microscopy, where an emphasis is paid to a uniform coverage of the field of view by illuminating light, in this case, the illumination and detection optics are focused on the same diffraction-limited spot with the aim to suppress significantly the out-of-focus light. Hence, the measurements are carried out discretely at the individual points forming an a priori-defined experimental grid. This grid represents a cross-section of the studied object. According to the specific arrangements there are three basic groups of the confocal microscopes differing in their method of scanning: laser scanning, spinning disk, and hybrid scanning (Elliott 2020).

The original Minsky configuration (Minsky 1957) holds the optics fixed and the microscope stage moved following the measurement grid. The indisputable advantage of this procedure is in preservation of identical optical properties during the whole

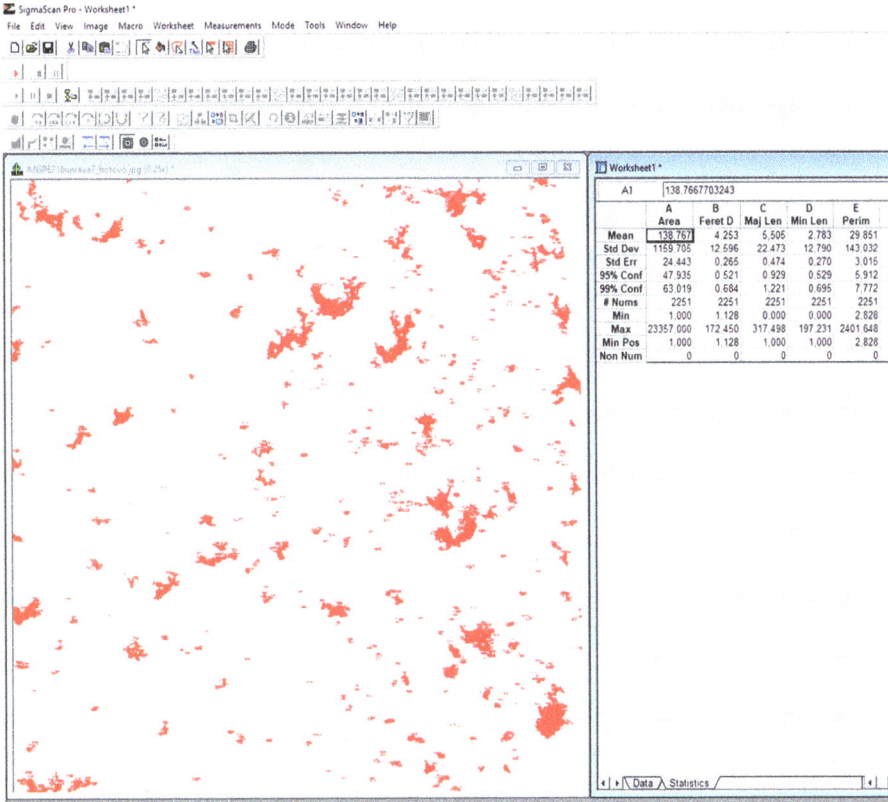

Fig. 4.31: Characterisation of the aggregates in Fig. 4.30 using the parameters: area, Feret diameter, maximum and minimum lengths, and perimeter (outputs from the software SigmaScan (Systat Software Inc.)).

measurement. On the other hand, it significantly increases the demands concerning the mechanical precision of the stage, and moreover, a motion can contribute to a re-arrangement of the studied samples, and the whole process is time-consuming in comparison to the present amply-used beam scanning with a stationary sample (see Fig. 4.32, right).

As an interstep between the classical light microscopy and scanning and transmission electron microscopy, the confocal laser scanning microscopy exhibits pros and cons with both limiting procedures (Wright et al. 1993; Bremer et al. 1993; Tata and Raj 1998; Paddock 2000; Teng et al. 2020):

– a probe can be located in hundreds of micrometres or in millimetres from the sample surface and not in nanometres as in the case of electron microscopes; there is also no need to measure in vacuum and sample preparation is much simpler; acquisition costs are moderate including customer service;

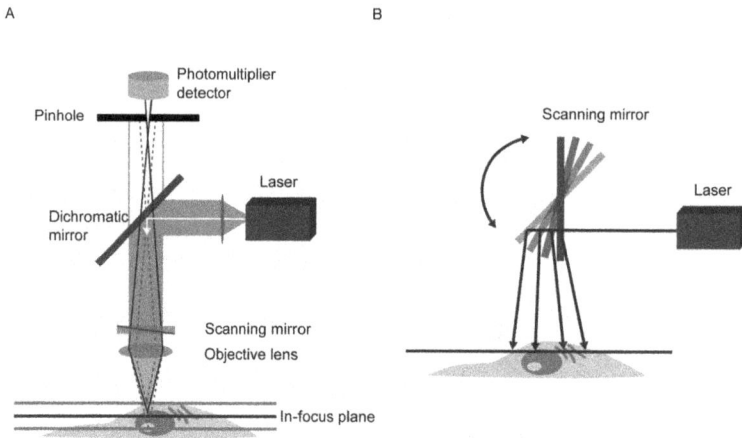

Fig. 4.32: A scheme of the confocal microscope (left) and sweeping the excitation light across the sample by scanning mirrors (right) (taken from Elliott 2020).

– in contrast to light conventional microscopy, light not coming from the focal plane is rejected, which enables to perform optical slicing and construction of 3D images.

The non-invasive CSLM method provides reconstruction of 3D structures (optical sectioning) within an analysed sample. It is achieved in more or less natural conditions (no geometrical sectioning manipulating with a sample is required). The accuracy of measurement is relatively precise (~0.2 µm laterally and ~0.6 µm axially, though in practice that is not always achieved, Elliott 2020). If a 3D reconstruction is created by stacking of the well-defined successive focal optical sections, a relatively thin thickness of the individual scanned layer (cross-sectional plane, shallow depth of field) should be compared with particle sizes.

The crucial shortcoming of the CSLM method is connected with a time interval required for capturing the measured data. Beam scanning with a fixed sample accelerates the measurements (nevertheless, slower scans provide a better signal-to-noise ratio resulting in better contrast). However, 3D reconstruction requires measurements in many cross-sectional planes. Using this method in determining a fractal dimension of coagulated aggregates, it is necessary to apply a fractal principle to a description of the studied floc aggregates. The fractal approach is based on the assumption of self-similarity; in other words, it is invariant whether this process is analysed from the first step or from any other one. If we project this property into the CSLM method then we can suppose that the intersections of the individual focused layers with the studied aggregates exhibit an identical fractal dimension. Hence, a measurement of only one layer is sufficient. This consideration reflects in a substantial reduction of time requirements.

Following Mandelbrot (1982), it is possible to relate a fractal dimension of the object O with a fractal dimension of the focused plane P (layer). The term 'codimension' is defined in the classical 3D Euclidean geometry as a dimension of the remaining space RS to the object O:

$$\dim O + \dim RS = 3 \Rightarrow \operatorname{codim} O \stackrel{def}{=} \dim RS. \tag{4.48}$$

From the following relation expressing the relation between a codimension of the intersection with the codimensions of the individual geometrical forms

$$\operatorname{codim} (O \cap P) = \operatorname{codim} O + \operatorname{codim} P, \tag{4.49}$$

we obtain

$$3 - \dim (O \cap P) = (3 - \dim O) + (3 - \dim P) \tag{4.50}$$

The dimension of the Euclidean plane P equals 2, and the dimensions $\dim (O \cap P)$ and $\dim O$ represent the fractal dimensions of the intersection region $D_{O \cap P}$ and the whole object D_f, respectively. After inserting in rel. (4.50) we obtain the relation

$$D_f = D_{O \cap P} + 1. \tag{4.51}$$

Using this procedure, it can be widened – in comparison with the box counting method – a description of fractal dependencies between various quantities (including mass fractal dimension) as carried out, for instance, in Thill et al. (1998).

4.4.2.2.3 Focused beam reflectance method

A number of particles forming the floc structure, their sizes, and spatial arrangement represent the basic characteristics for a description of aggregate behaviour. A technique of focused beam reflectance measurement (FBRM) contributes to experimental determination of these characteristics.

The principle of the FBRM method is shown in Fig. 4.33. A cylindrical probe immersed in a suspension is ended with a sapphire window (of a diameter ~ 12 mm) outside of which the infrared laser beam is focused. The rotating focused laser beam traces a circular path around the window circumference at a fixed speed of approximately 2 m s^{-1} or higher (up to ~6 m s^{-1}). The substantial point is that in this relation a particle velocity can be neglected. The focused beam gradually intersects the particles passing the window surface (see Fig. 4.34). The spot is not naturally ideal but exhibits approximately elliptical shape (~2 × 0.8 μm), where the lower dimension follows the scanning direction, thus providing maximum illumination. Nevertheless, the spot area depends on the focal position and the size and distance of the particle from the focal position.

Laser light impact induces backscattering of light to a detector positioned at 180° to the laser diode. The received intensity varies with the mutual location of the laser beam and primary particle surface (for a spherical particle, see Fig. 4.35). Backscatter-

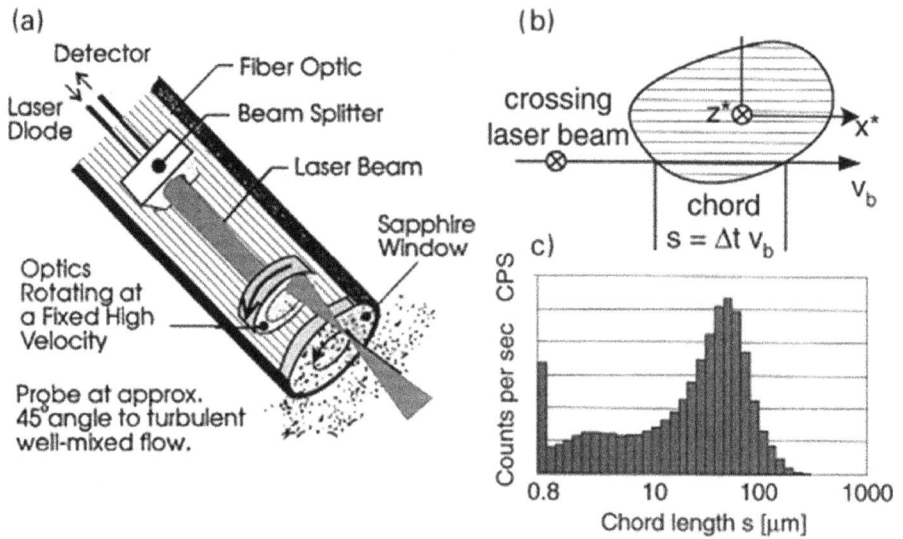

Fig. 4.33: A sketch of (a) the FBRM probe, (b) chord measurement (the laser beam direction perpendicular to the shaded area, crossing time Δt and the beam velocity v_b), and (c) histogram of the chord length counts (taken from Ruf et al. 2000, Fig. 1).

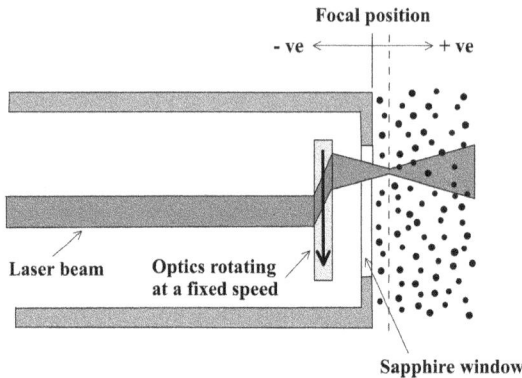

Fig. 4.34: A sapphire window with a rotating laser beam (taken from Heath et al. 2002, Fig. 1).

ing lasts till the laser beam leaves a particle surface where the path connecting both particle edges – called a chord – is terminated by both (initial and terminal) ends.

The FBRM optics collects backscattered light. This information is converted into an electronic signal (see Fig. 4.36). The time period of backscattered signal is known as well as the fixed rotational speed of a focused laser beam; hence, their multiplication provides information about the length of the chord. That is why the term FBRM is sometimes replaced by the term chord length distribution.

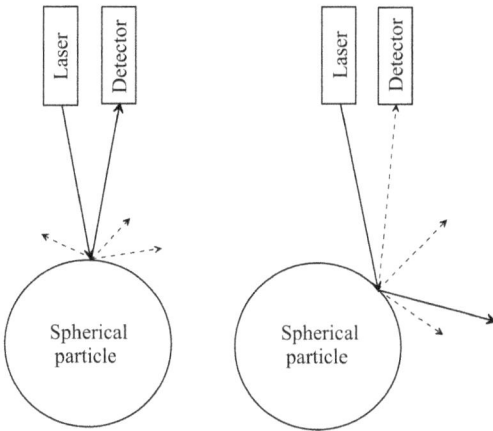

Fig. 4.35: Dependence of laser beam reflectance on a mutual position of the focused laser beam and particle surface: dominantly specular (left) and (b) diffusive reflectance (right) (taken from Kovalsky and Bushell 2005, Fig. 3).

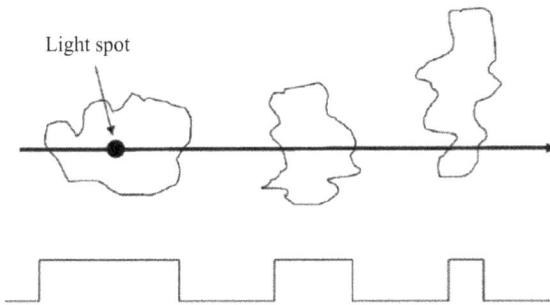

Fig. 4.36: An analogue-to-digital conversion of the FBRM measurements providing particle chord lengths (taken from Gregory 2009, Fig. 10).

The FBRM technique exhibits a series of advantages and disadvantages. Undoubtedly positive contribution is in a possibility to apply this method in situ. This means that – contrasting to some other methods – there is no necessity of sampling the studied suspended material and to expose it to shear damage due to pumping and turbulence. Further advantages include ease of use, little maintenance, and calibration requirements. Concentration is not restricted only to a dilute regime but the method is efficient up to 20% or more (concentration vs. chord measurements analysed in Yu and Erickson 2008).

The area of a focused laser beam spot governs the applicability of this method for particle sizes exceeding approximately 0.5 or 1 µm. This also generates the sampling frequency (approximately 2 MHz when a sample of 1 µm is taken per second under a rotational speed 2 m s^{-1}). Really sophisticated problem is created by a recalculation of chord lengths to a length parameter representing the particles. The easier problem is represented by spherical particles and conversion of the statistical-based (thousands of measurements taken in tens of seconds) chord length distribution to a determination of particle radius. Much more complicated problem appears for non-spherical particles (Heath et al. 2002; Ruf et al. 2000; Li and Wilkinson 2005; Ferreira et al. 2020).

Particle counting and evaluation of particle sizes provided by the FBRM technique contribute to fractal characterisation of aggregate morphology as presented by Kovalsky and Bushell (2005) and Lopez-Exposito et al. (2019). Kovalsky and Bushell (2005) used FBRM results for the determination of mass fractal dimension. Lopez-Exposito et al. (2019) applied a number of primary particles in a floc to calculate fractal dimension and analysed the relation between the values of fractal dimension and probability of intersections of the particles in corresponding aggregates with the laser beam.

4.4.2.2.4 Light scattering transmission – turbidity

Turbidity denotes a degree of haziness of a fluid caused by large numbers of microscopic invisible (by a naked eye) particles and represents one of the key factors in evaluating water quality. As a measure of water clarity, turbidity depends on intensity of absorption and scattering of passing light resulting from a presence of suspended particles. Scattering, that is, radiation deviating from the incident direction, is influenced by various factors such as particles shape and size, wavelengths of impinging light beams, and their relation to a characteristic dimension of the particles, refractive index, and inhomogeneity of the sample.

As a key factor of water clarity, a series of techniques have been developed for measurement of turbidity (see Sadar 1998; Omar and MatJafri 2009; Kitchener et al. 2017; Matos et al. 2024). These techniques differ not only in their accuracy and repeatability but also in the units describing a measure of turbidity. The classical approaches are based on evaluating of transmitted light (turbidimetry) and in principle are based on quantification of a ratio transmitted light energy vs. incident light energy. Light is not only transmitted or scattered but also absorbed depending on the wavelength of the incident light.

However, more accurate results can be achieved by the so-called nephelometric technique (nephelometry). Nephelometry quantifies the amount of light scattered away from the direction of the incident light. The scattered light is less affected by the size of suspended particles. By convention a scattered light detector is located perpendicularly (90°) (see Fig. 4.37).

Complexity of this topic is underlined by the fact that a scale of turbidity does not start from 0 as is common for practically all measured parameters. Incident light beams pass the sample of pure water relatively undisturbed; however, the molecules themselves participate in light scattering and the value attains approximately 0.012 nephelometric transmission unit (NTU).

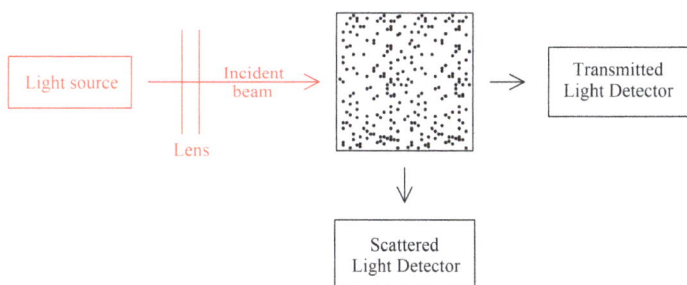

Fig. 4.37: The locations of transmitted and scattered light detectors in relation to a direction of an incident beam.

A definition of NTU unit is based on a formazin standard which preparation is carried out in three consequent steps (US EPA method 180.1, https://www.epa.gov›files›method_180-1_1993):

–   1.00 g hydrazine sulphate is dissolved in reagent water and diluted to 100 mL,
–   10.00 g hexamethylenetetramine is dissolved in reagent water and diluted to 100 mL,
–   5.0 mL of each solution is mixed and left for 24 h at 25 ± 3 °C and then diluted with reagent water up to 100 mL.

Finally, 10 mL of this stock mixed with 90 mL of reagent water represents a standard which turbidity corresponds to 40 NTU.

Formazin nephelometric unit (FNU) complies with the European ISO method 7027, the European drinking-water protocol, and the following relation is valid: 1 NTU = 1 FNU. A notable difference between the two standards, EPA method 180.1 and ISO method 7027, is in the applied wavelength of light beams 400–600 nm and 860 nm, respectively. The third turbidity standard GLI method 2 (based also on formazin standards, also denoted as four-beam modulated turbidimetry) is explicitly applied for the determination of turbidity in drinking water and uses a different experimental arrangement (see Fig. 4.37). The 'empty' direction is occupied by another light source and the measurement carries out in two phases (see Fig. 4.38). Initially, the first light source (Run No.1) impinges an incident light and the transmitted and scattered data are collected. Consequently, within a short time interval (approximately 0.5 s), the second light source (Run No.2) is used, and in this case, the transmitted and scattered detectors are swapped. Hence, four measurements are obtained using this nephelometric and attenuation-based method (infrared light of 860 nm wavelength as in ISO method 7027-2:2019 is applied). The turbidity value *Turb* is calculated from these four readings using the calibrated geometric mean

$$Turb = Cal_{slope} \sqrt{\frac{I_{1,scattered}}{I_{1,transmitted}} \times \frac{I_{2,scattered}}{I_{2,transmitted}}} - Cal_{shift}, \qquad (4.52)$$

where $I_{i,scattered}$ and $I_{i,transmitted}$ are the currents [mA] in the respective detectors when the i-light source is on and the other one off, $i = 1,2$; Cal_{slope} and Cal_{shift} are the calibration constants.

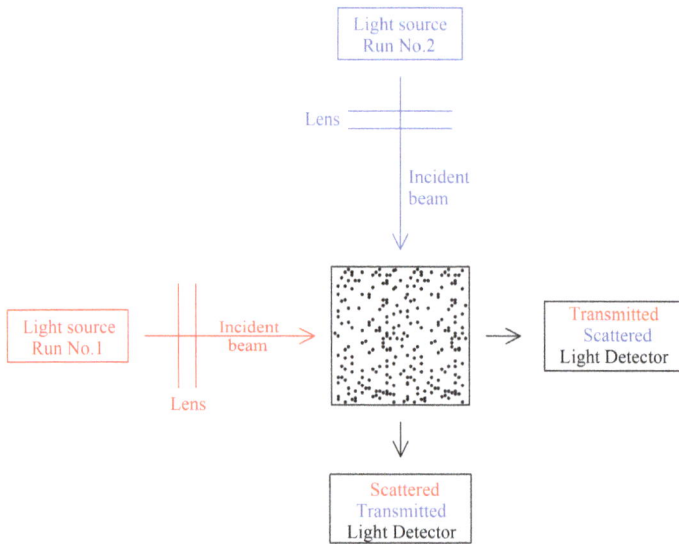

Fig. 4.38: The arrangement of light sources and detectors according to GLI method 2: turbidity.

Formazin (very poorly soluble in water) can be reproducibly prepared (deviation lower than 1%) and is composed of a wide range of particle shapes and sizes ranging from 0.1 to 10 mm. On the other hand, formazin exhibits various shortcomings: carcinogenicity, concentration uncertainty, poorer stability with a concentration decrease (lower than 1 h for NTU lower than 2), the resulting particle size distribution subjects to the preparation temperature, and uncertainties related to temperature fluctuations are of the order of 1.2% per °C (Buzoianu 2000). The decreasing stability can influence turbidity measurements of drinking water where the upper bound is set to 5 NTU. For completeness, the relation between NTU and suspended solids is as follows: 1 mg/L (1 ppm) is equivalent to 3 NTU. These shortcomings can question universality of turbidity determination based on formazin and other standards (e.g. a standard based on styrene divinylbenzene polymer).

Another possibility how to quantify turbidity is an orientation to purely optoelectronic characteristics as turbidity is not connected with a measurement of suspended particles in water but is governed by a measure of scattering effect of the particles on light. This can be documented on an example of refractive indices. The increasing ratio of the particle-to-water refractive indices results in more intensive scattering. Reversely, if this ratio approaches 1, then scattering vanishes. The complementary characteristic is represented by absorption of light by solute molecules. With increasing particle concentration, scattered light strikes more particles which cause an increase in light absorption. A crucial indicator participating in a ratio absorbed vs. scattered light is an absorption parameter. Its course for pure water in dependence on wavelength of the incident light is depicted in Pope and Fry (1997). First, with an increasing wavelength λ the absorption parameter approaches its minimum at $\lambda = 417.5$ nm and then dramatically increases.

Multi-wavelength turbidimetry (MWT) relates incident light intensity I_0 with the transmitted light intensity I (both dependent on the wavelength λ) by the relation

$$I(\lambda) = I_0(\lambda) \times \exp(-\tau(\lambda)/L) \tag{4.53}$$

or equivalently

$$\tau(\lambda) = L \times \ln\left(\frac{I_0(\lambda)}{I(\lambda)}\right), \tag{4.54}$$

where $\tau(\lambda)$ is the turbidity and L is the optical path (that should be lower than 0.1 m). The reducing factor $e^{-\tau(\lambda_0)/L}$ is composed of two contributions: absorption and scattering. For samples not exhibiting absorption when the attenuation is caused only by scattering, the MWT technique represents a truly scattering procedure.

The power relation

$$\tau(\lambda) \propto \lambda^\beta \tag{4.55}$$

between the turbidity and the wavelength was theoretically derived by Doty and Steiner (1950) and Horne (1987). As stated in Senesi et al. (1997), the system should contain monodisperse particles, almost identical in size and shape and large enough to scatter independently. To avoid multiple scattering effects and particle interactions, the use of dilute suspensions is required. Horne (1987) introduced a relation

$$\beta = 4.2 + \frac{\mathrm{dlog}\tau(\lambda)}{\mathrm{dlog}\lambda}. \tag{4.56}$$

The value of parameter β determines a fractal dimension based on turbidity data. The value of $\beta < 3$ corresponds to mass fractal dimension. Within the interval $3 < \beta < 4$, a surface fractal equals to $6 - \beta$ (for details, see Senesi et al. 1997).

Another approach how to determine a fractal dimension from the turbidity measurements was introduced by Cheng et al. (2010). Their approach was inspired by the procedure introduced in Yukselen and Gregory (2002) who applied the so-called flocculation index FI (see Eisenlauer and Horn 1985; Gregory and Nelson 1986; and Fig. 4.39) to a determination of strength and recovery factors of flocs

$$\text{Strength factor} = 100 \times \frac{\text{FI}_{\text{br}}}{\text{FI}_{\text{in}}} \ [\%], \tag{4.57}$$

$$\text{Recovery factor} = 100 \times \frac{\text{FI}_{\text{re}} - \text{FI}_{\text{br}}}{\text{FI}_{\text{in}} - \text{FI}_{\text{br}}} \ [\%], \tag{4.58}$$

where FI_{in} and FI_{re} denote the maximum FI values of the flocs during the initial and re-formation stages, respectively; FI_{br} corresponds to the value when the flocs are broken.

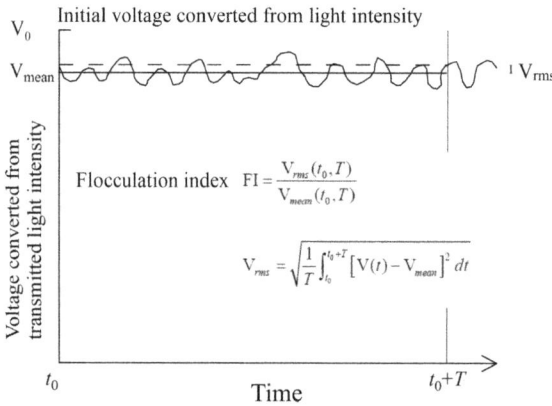

Fig. 4.39: A process of the determination of the flocculation index.

Cheng et al. (2010) re-formulated the preceding relations to the form

$$\text{Strength factor} = 100 \times \frac{\text{SD}_{\text{br}}}{\text{SD}_{\text{in}}} \ [\%], \tag{4.59}$$

$$\text{Recovery factor} = 100 \times \frac{\text{SD}_{\text{re}} - \text{SD}_{\text{br}}}{\text{SD}_{\text{in}} - \text{SD}_{\text{br}}} \ [\%], \tag{4.60}$$

where the flocculation indices are replaced by the standard deviations of turbidity measurements in the respective stages (a time interval T was set to 120 s). They applied an on-line continuous turbidity monitoring system in jar test measurements using kaolin suspension in water. A determination of the fractal dimension was based on the relation area vs. floc perimeter using a digital camera. Their data is processed in Fig. 4.40.

Finally, it should be noted that the turbidity itself is not dependent on temperature; however, its measurement – if submersible sensors with an LED light source are used – yes. This is caused by a dependence of the electrical output of the sensors on temperature changes. It is necessary to take these changes into account and compensate.

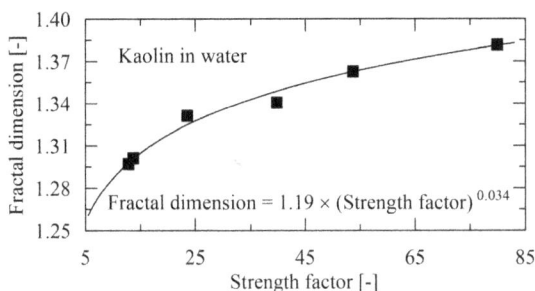

Fig. 4.40: A dependence of fractal dimension on the strength factor determined by standard deviations of turbidity measurements (data taken from Cheng et al. 2010).

4.4.2.2.5 Static light scattering

As shown above light transmission and nephelometric measurements (scattering at the angle 90°) exhibit very positive characteristics: non-invasiveness and providing information practically in real time. These attributes are valid for all scattering (X-rays, neutron, light) methods which usage differs according to the individual dimensions. From the position of fractal analysis it implies that the derived fractal dimensions evaluate real (not disturbed) suspensions under study.

In the case of the scattered light, its intensity subjects to a series of fundamental parameters (Gregory 2009):
- particle size as a radius r,
- light wavelength λ (in hundreds of nm),
- a ratio $x = 2\pi r/\lambda$ relating both preceding parameters (=a perimeter of the particle/wavelength),
- particle refractive index relative to the suspension medium m (=(refractive index of the scatterers)/(refractive index of a carrier liquid)),
- scattering angle θ ($\theta \in (0°, 90°)$ for forward scattering, $\theta \in (90°, 180°)$ for backscattering, the angles express a deviation from the direction of an incident beam.

Each of these parameters has to be restricted in deriving the description of the process of light scattering measurement as the interpretation of the scattered intensity pattern is complicated by the strong interaction of light and matter. The elastic scattering is one of these simplifications. It means that preserving the internal states of the scattering particles is supposed (no exciting of electrons, no creation of new particles). Hence, the total kinetic energy of the system is conserved. No (or negligible) absorption of light implies that the refractive indices attain only real values (or an imaginary part can be neglected).

In scattering, morphology of the particles plays a substantial role not only from the physical viewpoint but also simultaneously with the mathematical description of the whole process. This is the reason why a spherical shape is ordinarily taken into account. However, only the restriction of particle shape to a spherical geometry still does not provide a direct methodology how to derive physically relevant conclusions. The starting position significantly depends on the above-introduced ratio x (= $2\pi r/\lambda$).

Rayleigh scattering is based on the fulfilment of two conditions:
- the particles are much smaller than the wavelength of the incident light, i.e. $x \ll 1$;
- the particles are supposed to be optically soft; in other words, their refractive index is very close to the one of the carrier liquid, i.e. $|m - 1| \ll 1$.

Under these assumptions scattering is spatially isometric (equal scattering in all directions) and strongly dependent on the incident wavelength ($\sim\lambda^{-4}$) and particle size ($\sim r^6$).

The Mie theory deals with the particles in which radius is comparable or larger than a wavelength of the incident beam ($x \cong 1$). Spatial isometry is no longer valid (see Fig. 4.41). The interplay between particle sizes and wavelengths of the incident beam is roughly depicted in Fig. 4.42, dependence of scattered intensity on particle size is reduced from the sixth power to the square one ($\sim r^2$).

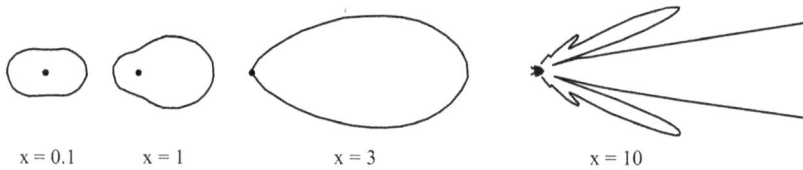

$x = 0.1$ $x = 1$ $x = 3$ $x = 10$

Fig. 4.41: Pronounced intensity of forward scattering for increasing size of particles, light incident from the left (adapted from Kitchener et al. 2017, Fig. 2).

The Rayleigh-Gans-Debye (RGD) theory supposes optically soft monodisperse particles which sizes are comparable with the incident wavelength and that the extent of multiple scattering can be neglected. Further, it is supposed that light is not reflected at the particle-medium boundary and is not attenuated within the particle. These assumptions are expressed by the following relations (Bushell et al. 2002; Tang et al. 2002)

$$|m - 1| \ll 1, \tag{4.61}$$

$$\frac{2\pi n}{\lambda} L |m - 1| \ll 1, \tag{4.62}$$

where m is the refractive index of dispersion medium and L is the characteristic length scale of particles. This also ensures that each small portion of a particle is subject to the same electromagnetic radiation conveyed by the incident light and scatters light as if isolated from the rest of the particle. If dispersion medium is water, then the values (at 20 °C) of the refractive index n are 1.34 for 350 nm $< \lambda \leq$ 500 nm and 1.33

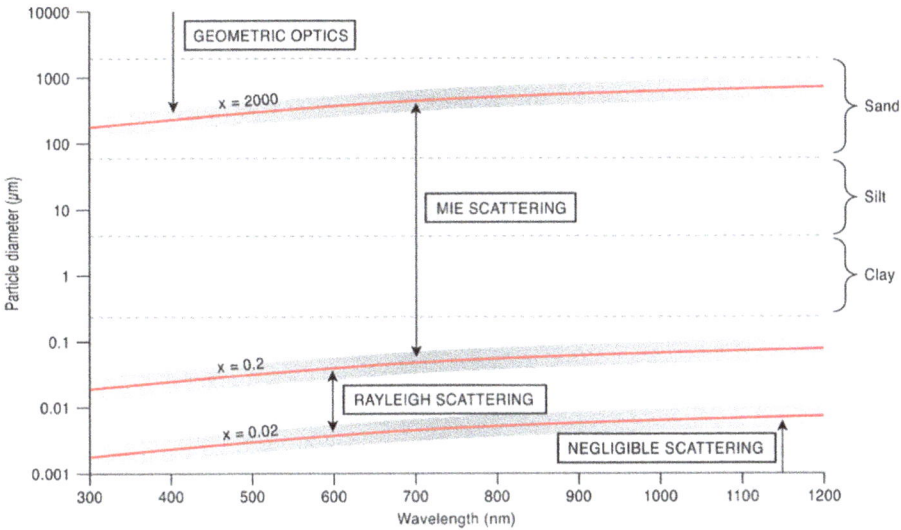

Fig. 4.42: Approximate borders among light scattering theory regimes as a function of particle diameter (2r) and wavelength of incident light λ. Particle size bands sorted according to the American Geophysical Union Sediment Classification System (taken from Kitchener et al. 2017, Fig. 4).

for 500 nm $< \lambda \leq 1,000$ nm (see https://en.wikipedia.org/wiki/Optical_properties_of_water_and_ice). The deviations of the refractive index do not exceed 0.1% within the temperature range (0 and 30 °C) (see https://www.engineersedge.com/physics/refraction_for_water_15690.htm).

The restriction (4.62) – not required in the Mie approach – was attenuated by Farias et al. (1996) who showed – using numerical experiments with aggregates up to several hundred primary particles – that within an accuracy up to 10% the strict demand $\ll 1$ can be substituted by much weaker ≤ 0.6.

The pilot notion in the RGD approximate method represents a scattering wave vector defined as a subtraction of the incident $(\vec{k_i})$ and scattered $(\vec{k_s})$ beam vectors, where for coherent elastic scattering, the relation $|\vec{k_i}| = |\vec{k_s}| = 2\pi n/\lambda$ is valid (see Fig. 4.43). Its magnitude is given by the relation

$$ q = |\vec{Q}| = \frac{4\pi n}{\lambda} \sin\left(\frac{\theta}{2}\right), \tag{4.63} $$

where θ is the angle at which the radiation is scattered (see e.g. Sorensen 2001; Bushell et al. 2002).

The scattered intensity $I(q)$ can be expressed by the product of the so-called primary particle form factor $P(q)$ and the interparticle structure factor $S(q)$

$$ I(q) \propto P(q) \times S(q). \tag{4.64} $$

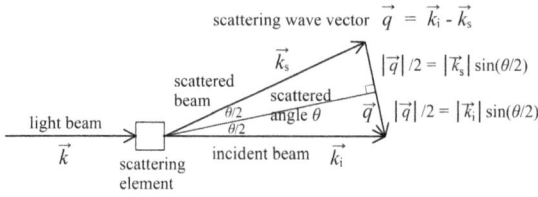

Fig. 4.43: A geometrical interpretation of the scattered wave vector.

The form factor $P(q)$ is related to the shape of the particles while the structure factor $S(q)$ characterises their spatial arrangement and describes the correlations between individual particles in an aggregate, assuming that there are no correlations between the aggregates. From the viewpoint of scattering, $P(q)$ describes the scattered intensity function from a single primary particle and $S(q)$ describes the additional scattered intensity due to the spatial correlation between the particles in the aggregate.

For spherical particles the form factor $P(q)$ is given by the relation (see https://lsinstruments.ch/en/theory/static-light-scattering-sls/theory)

$$P(q) = \left[\frac{3}{(qr)^3} (\sin(qr) - qr\cos(qr)) \right]^2 . \tag{4.65}$$

It is apparent that for the product qr approaching 0 ($qr \ll 1$) the form factor attains a value 1 (see Fig. 4.44). For larger values of qr, a course of the form factor is influenced by the interparticle interference effects and nullifies at the points $\tan(qr) = qr$.

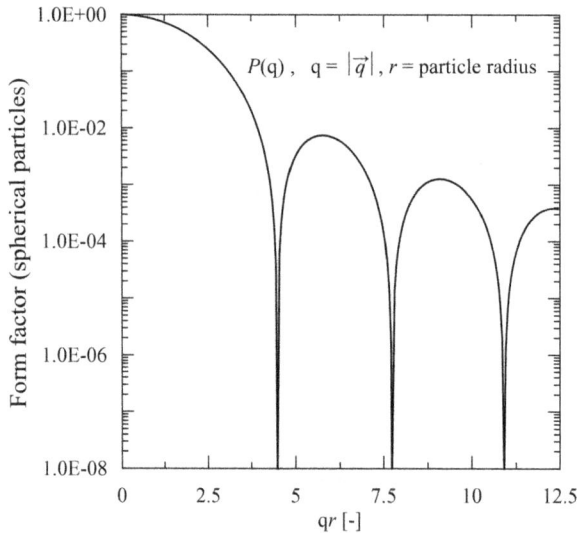

Fig. 4.44: A course of the form factor $P(q)$ simplified for the spherical particles (eq. (4.65)).

The initial plateau of the form factor can be very well approximated by the simpler relation (the so-called Guinier approximation, Sorensen et al. 1995)

$$P(q) \simeq 1 - \frac{(qR_g)^2}{3},$$ (4.66)

where R_g is the radius of gyration. The radius of gyration represents a distance to an axis of rotation with the moment of inertia exactly the same as the particle (aggregate) actual distribution of mass. Figure 4.45 indicates a proportion of the radius of gyration to the whole aggregate. For a single spherical particle, the relation $R_g = \sqrt{3/5}\,r$ holds.

Depicting the behaviour of time-averaged scattered intensity $I(q)$ (see Fig. 4.46), we can see qualitative changes in three regions (see Bushell et al. 2002). The first one ($qR_g < 1$) – the Guinier regime (small angle dependence of scattered intensity) – is governed by the Guinier approximation. The Porod regime ($qr \gg 1$) is governed by the Porod law

$$I(q) \propto q^{-4}.$$ (4.67)

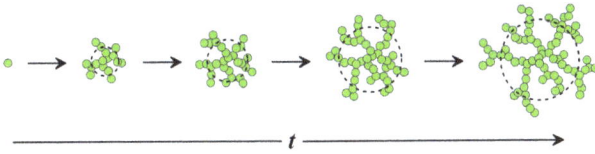

Fig. 4.45: Identity of the moment of inertia of the radius of gyration and the whole aggregate (taken from Ferri et al. 2011, Fig. 1).

The intermediate region characterised by $1/R_g \ll q \ll 1/r$ enables due to a relation

$$I(q) \propto q^{-D_f}$$ (4.68)

to determine a fractal dimension of the aggregates. This relation is based on rel. (4.64); as in this region, the following expressions hold (Biggs et al. 2000)

$$P(q) \sim 1, S(q) \propto q^{-D_f}.$$ (4.69)

Relation (4.68) is often used for the determination of the fractal dimension (e.g. Sorensen 2001; Bushell et al. 2002; Lin et al. 2003; Yang et al. 2014; and references therein). However, processing of the experimental data faces possible obstacles.

An accuracy of the calculated fractal dimension of the aggregates can be worsened by inadequate particle concentration. Low concentration can result in a non-representative scattering pattern; on the other hand higher concentration can cause unwanted multiple scattering (see Greenwood et al. 2007). The presence of multiple

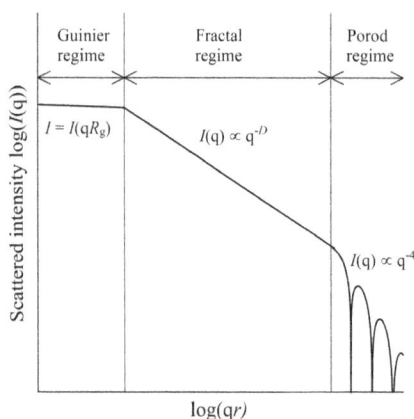

Fig. 4.46: Three distinct scattering regimes according to behaviour of the scattered intensity in dependence on qr (monodisperse spherical particles are supposed). The transition between the Guinier and fractal regimes is more complicated than sketched (see Jullien 1992).

scattering invalidates the RGD conditions upon which the analysis is based (Bushell et al. 2002).

Another problem is that the assumptions in the RGD theory are not restrictive to the concentration only but also with respect to the particles (monodisperse spherical). However, even if these two assumptions of the RGD theory are violated, the power dependence of scattered intensity on the magnitude of scattered vector may still exist. In this case the term 'fractal dimension' is replaced by the term 'scattering exponent' (Stone et al. 2002).

In detail analysis the passage from the Guinier regime to the fractal one exhibits an inflection point that should be eliminated in determining a fractal dimension (Jullien 1992). Hence, the requirement $qR_g > 1$ need not to be sufficient and the value 1 should be increased on the case-by-case basis.

Any procedure determining the fractal dimension has its pros and cons; however, static light scattering as a non-invasive technique does not destroy aggregate morphology which is crucial in determining the fractal dimension of flocs.

4.4.3 Fractal dimension based on settling velocity

Sedimentation along with filtering are two processes used in water and waste water treatment which efficiency is based on successful coagulation and flocculation of primary particles contained in the raw media. This evokes a natural question whether a determination of the fractal dimension of created aggregates can be based on knowledge of their settling velocity.

(Not)surprisingly the answer is not trivial. The traditional Stokes' law describing the drag force (acting on the interface between the carrier liquid and the particle) resisting the fall of small spherical particles through a carrier liquid is of the form

$$F_D = 6\pi r_p \eta v_s,\tag{4.70}$$

where r_p is the radius of the spherical particle, η is the dynamic viscosity, and v_s represents the settling velocity. Using the Archimedes' principle, the buoyant force reduces the force acting downwards to the form

$$F_{dw} = \frac{4}{3}\pi r_p^3 \left(\rho_p - \rho_l\right)g,\tag{4.71}$$

where ρ_p is the density of the particle, ρ_l is the density of the carrier liquid, and g is the acceleration of gravity. Equating both forces, an expression for the settling velocity is ($d_p = 2r_p$)

$$v_s = \frac{(\rho_p - \rho_l)g}{18\eta}d_p^2.\tag{4.72}$$

The problem is that – in contrast to coagulation and flocculation – the following assumptions are supposed:
(a1) laminar flow (low Reynolds numbers $\mathrm{Re} = \rho_l v_s d_p/\eta$),
(a2) negligible inertial effects ($\mathrm{Re} < 1$),
(a3) spherical shape of the non-porous particles,
(a4) homogeneous material (uniform structure),
(a5) smooth particle-liquid interface,
(a6) no interference among particles.

Each of these assumptions is in a strong contradiction with practical aspects of coagulation and flocculation:
(b1) mixing in sedimentation tanks does not ensure laminar regime,
(b2) very low values of Re cannot be fulfilled,
(b3) in principle, forming aggregates are porous and non-spherical,
(b4) any medium is non-uniform,
(b5) encapsulation appearing in aggregate morphology does not contribute to smooth interface,
(b6) coagulation and flocculation exhibit sticking (destabilisation) of particles and aggregates.

The analogous situation is with the classical expression for the drag coefficient C_D which represents a dimensionless quantification of the resistance of a particle (aggregate) in a fluid environment. Using the drag coefficient, the drag force can be expressed in the form

$$F_D = C_D \times \frac{\rho_l v_s^2}{2} \times A\tag{4.73}$$

where A is the projected area of a spherical particle, i.e. $\pi d_p^2/4$. Comparing rels. (4.70) and (4.73) we get the so-called Stokes' law

$$C_D = \frac{24}{Re},\qquad (4.74)$$

where the Reynolds number is introduced as

$$Re = \frac{\rho_1 v_s d_p}{\eta}.\qquad (4.75)$$

As rel. (4.70) was used in a derivation of the relations, a list of the assumptions (a1)–(a6) also passes to the position of C_D.

As expectable – due to discrepancies between the points (a1)–(a6) and (b1) –(b6) – the theoretically derived above-introduced relations describing behaviour of non-porous spherical particles in the surrounding media are not applicable to the description of coagulation and flocculation processes in practice (Johnson et al. 1996; Li and Logan 2000; Bushell et al. 2002; Maggi 2007; Vahedi and Gorczyca 2011, 2012, 2014; Chakraborti and Atkinson 2020; Moruzzi et al. 2020b; and references therein).

The common approach how to suppress partially the strict assumptions (a1)–(a6) with an inclination to more realistic situation (b1)–(b6) has been reflected in the generalisation of the basic functional structures for v_s (rel. (4.72)) and C_D (rel. (4.74)). The emphasis has been paid on three aspects:
– Functional expansion of the terms depending on the dimensionless Reynolds number Re relating inertial and viscous forces; this ensures applicability to larger values of Re (in comparison with the condition Re < 1);
– Introducing the so-called shape parameter accommodating the resulting expressions for specific non-spherical particles;
– Respecting permeability of the particles substantially changing flow resistance of the surrounding media. Permeability is a measure of the ability of porous material to allow fluids to pass through it. Apart from porosity, it also subjects to the shapes of the pores and their level of connectedness.

Schiller and Naumann (1933) proposed the commonly used empirical relation generalising the Stokes' law (rel. (4.74))

$$C_D = \frac{24}{Re}\left(1 + 0.150\, Re^{0.687}\right),\qquad (4.76)$$

where the binomial factor approximates already non-negligible inertial part (for Re < 800). For lower values of Re (Re < 0.1) the deviation from the Stokes' law decreases below 3%.

Let us denote Ω a ratio comparing resistances of the permeable and impermeable spheres of the identical radia r. If we relate the radius r with permeability k of a spherical particle, then Neale et al. (1973) proposed the following relation

$$\Omega \;=\; \frac{2\beta^2\left(1-(\tanh\beta/\beta)\right)}{2\beta^2+3\left(1-(\tanh\beta/\beta)\right)}, \tag{4.77}$$

where

$$\beta \;=\; \frac{r}{\sqrt{k}}. \tag{4.78}$$

For $\beta \geq 5$ ($\tanh\beta \approx 1$) the above relation can be simplified to

$$\Omega \;=\; \frac{2\beta^2(\beta-1)}{2\beta^3+3(\beta-1)}. \tag{4.79}$$

A ratio Ω can be used as a proportional factor in expressing the drag force and hence implemented in the expression for the settling velocity as used by Masliyah and Polikar (1980). They also adapted the functional dependence on the Reynolds number with the aim to widen its range

$$C_{\mathrm{D}} \;=\; \frac{24\Omega}{\mathrm{Re}} \times \left(1+0.1315\,\mathrm{Re}^{0.82-0.05\log(\mathrm{Re})}\right) \text{ for } 0.1 < \mathrm{Re} \leq 7 \tag{4.80}$$

and

$$C_{\mathrm{D}} \;=\; \frac{24\Omega}{\mathrm{Re}} \times \left(1+0.0853\,\mathrm{Re}^{1.093-0.105\log(\mathrm{Re})}\right) \text{ for } 7 < \mathrm{Re} < 120. \tag{4.81}$$

Based on the experimental measurements for their materials, these relations are valid within the range $15 < \beta < 33$.

For a determination of the fractal dimension F_{D}, Kranenburg (1994) proposed (under the assumption of monodisperse spherical primary particles) the following expression

$$\rho_{\mathrm{f}}-\rho_{\mathrm{l}} \;=\; \left(\rho_{\mathrm{p}}-\rho_{\mathrm{l}}\right)\left(\frac{d_{\mathrm{f}}}{d_{\mathrm{p}}}\right)^{F_{\mathrm{D}}-3}, \tag{4.82}$$

where ρ_{f}, ρ_{l}, and ρ_{p} are the densities of aggregate, carrier liquid, and primary particles, respectively, and d_{f} and d_{p} are the equivalent floc and primary particles diameters, respectively.

Khelifa and Hill (2006) introduced a fractal dimension variable with the size of flocs (d_{i} are the diameters of the individual primary particles) and proposed the relation

$$\rho_f - \rho_l = (\rho_p - \rho_l) \left(\frac{d_f}{d_{50}}\right)^{F_D - 3} \frac{\frac{1}{k}\sum\limits_{i=1}^{k} d_i^3}{\left(\frac{1}{k}\sum\limits_{i=1}^{k} d_i^{F_D}\right)^{3/F_D}}, \tag{4.83}$$

where d_{50} represents a median of the primary particles diameters (for details, see Khelifa and Hill 2006). Consequently, they obtained the relation for the settling velocity

$$v_s = \frac{\theta(\rho_p - \rho_l)g}{18\eta\left(1 + 0.15 Re^{0.687}\right)} \frac{d_f^{F_D - 1}}{d_{50}^{F_D - 3}} \frac{\frac{1}{k}\sum\limits_{i=1}^{k} d_i^3}{\left(\frac{1}{k}\sum\limits_{i=1}^{k} d_i^{F_D}\right)^{3/F_D}}, \tag{4.84}$$

where θ is the floc shape parameter. This relation projects onto the Stokes' formula (4.71) for spherical ($\theta = 1$) and non-porous (constant $F = 3$) flocs and monosized particles ($d_p = d_i = d_{50}$).

Moruzzi et al. (2020a,b) used the formulas for porosity and floc concentration depending on the fractal dimension to a proposal of the expression for the settling velocity containing also a shape parameter.

The detailed reviews on published relations of the settling velocity v_s, and drag coefficient C_D are presented in Tang and Raper (2002), Cheng (2009), Hongli et al. (2015), Dey et al. (2019), Kramer et al. (2021).

4.4.4 Fractal dimension in relation to pH

The equilibrium molar concentration (mol/L) of the hydrogen ion H^+ in the solution indicates acidity or basicity of liquid solutions, which is expressed through so-called hydrogen ion concentration (potential of hydrogen)

$$pH = -\log([H^+]). \tag{4.85}$$

This relation, simplifying a real hydrogen ion concentration [H^+] (in the range 1 and 10^{-14} gram-equivalents per litre) and using a common logarithm, describes power-law changes in [H^+]. It means that shifting a value of pH by 1 is reflected by a change of one order in hydrogen ion concentration.

A setting of the correct pH value is a necessary condition for optimal efficiency of the process of coagulation (Gregory and Carlson 2003; Qin et al. 2006; Sillanpää et al. 2018; Naceradska et al. 2019; Ren et al. 2022; Pivokonsky et al. 2024; and references therein). This value is often close to neutral pure water (pH = 7); however, it is necessary to have in mind a dramatic change of hydrogen ions H^+ both in the acidic case (e.g. for pH = 6) or in the basic (alkaline) case (e.g. for pH = 8) as emphasised above.

The values of pH along with a choice of coagulant and its dosage represent three factors dominantly governing the whole process of coagulation including a significant participation in a reduction of the repulsive potential of electrical double layer of colloids as the first step. Not surprisingly there is a strong relation between used pH values (acidity or alkalinity) and the corresponding fractal values of the resulting flocs. These relations subject the used coagulants and their dosage.

The pH-fractal dependencies were analysed for both traditional coagulants (Al- and Fe-based salts) as well as for not so frequently used coagulants as for instance titanium tetrachloride. Loosely speaking, it is possible to say that efficiency of the process of coagulation increases with compactness of the formed flocs (better sedimentation) which is indicated by higher values of the fractal dimension. Lower values are characteristic for more porous floc structures.

Yu et al. (2014) investigated the formation, breakage, and re-growth of flocs using aluminium sulphate octadecahydrate $(Al_2(SO_4)_3 \cdot 18H_2O)$ as a coagulant at acidic pH. The fractal dimension was calculated with the help of a relation between total scattered light intensity and the scattering vector Q using small angle laser light scattering. They showed an apparent increase of fractal dimension between pH = 5.2 and 6.0 (see Fig. 4.47). Furthermore, the strong correlation between fractal dimension and strength of flocs was presented.

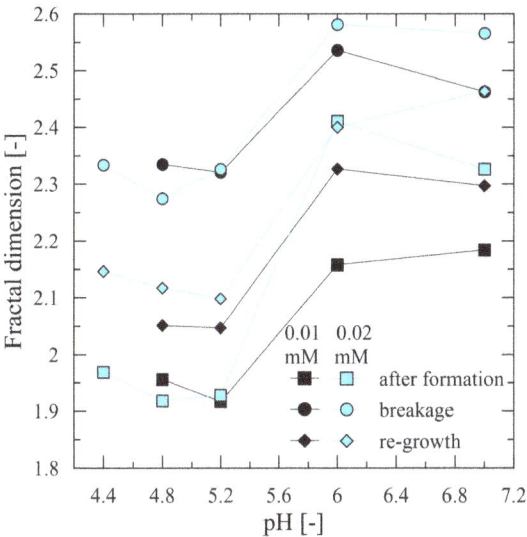

Fig. 4.47: A comparison of fractal dimensions for two doses of coagulant (aluminium sulphate octadecahydrate $(Al_2(SO_4)_3 \cdot 18H_2O)$) and at three stages of aggregate formation: after initial formation, breakage, and re-growth (data taken from Yu et al. 2014, Fig. 4).

Wang et al. (2011b) treated humic acids (HA) diluted with tap water using as a coagulant polyferric chloride (PFC, iron(III) chloride). Figure 4.48 indicates an optimal choice of a coagulant dosage and pH. The most compact PFC-HA flocs were produced at pH = 6 with a PFC dosage of 4.04×10^{-4} mol/L (Fe^{3+}); on the other hand the loosest flocs appeared at pH = 7 with a PFC dosage of 1.01×10^{-4} mol/L (Fe^{3+}). A determination of fractal dimension followed the procedure in Sokołowska and Sokołowski (1999) combining the Frenkel-Halsey-Hill equations and thermodynamic model.

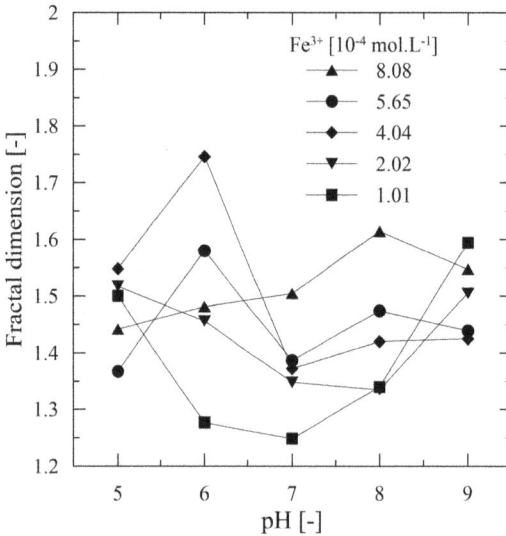

Fig. 4.48: A comparison of fractal dimensions of PFC-HA flocs for five different concentrations of a coagulant PFC (data taken from Wang et al. 2011b, Fig. 6b).

Senesi et al. (1996) measured a fractal dimension of HA in dependence on time at six different pH values. Fractal dimension was determined from the relation between turbidity and a wavelength. As depicted in Fig. 4.49, the initial humic particles exhibiting nearly space-filled structure (fractal dimension close to 3 for pH = 3) gradually (after 16 h) converted to looser objects. The same authors continued in this topic in Senesi et al. (1997).

A relation between fractal dimension and pH was also studied for other coagulants such as poly(acrylamide-acryloyloxyethyl trimethyl ammonium chloride-butyl acrylate) (Sun et al. 2015b). Xu et al. (2014) compared fractal behaviour with other characteristics for four coagulants: polyaluminium chloride (PACl), nano-Al_{13}, aluminium sulphate octadecahydrate $Al_2(SO_4)_3.18H_2O$, and nano-Al_2O_3 + aluminium sulphate octadecahydrate $Al_2(SO_4)_3.18H_2O$.

Water treatment residuals in the form of sludge water in sedimentation tank and filter backwash represent a permanent problem accompanying a production of pota-

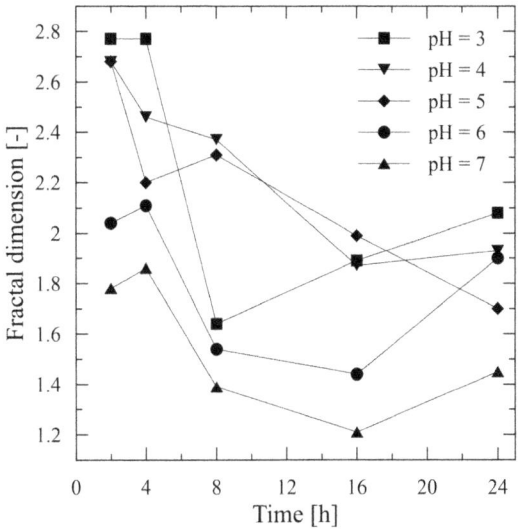

Fig. 4.49: Time courses of fractal dimensions of humic acid in distilled water in a dependence on pH values (data taken from Senesi et al. 1996, Table 1).

ble water in the drinking-water treatment plants. To reduce transport cost, sludge has to be flocculated and dewatered. The only exception is probably usage of titanium tetrachloride (TiCl$_4$) as a coagulant. In this case sludge water contains a valuable photocatalyst TiO$_2$ (Zhao et al. 2014). Sun et al. (2015a) used poly(acrylamide-acryloyloxyethyl trimethyl ammonium chloride-butyl acrylate) as a coagulant in three versions differing in cationic degree and intrinsic viscosity. The courses of fractal dimension in dependence on pH are depicted in Fig. 4.50 indicating that an optimal choice of pH in sludge water is approximately pH = 7. Fractal dimension was determined using a relation: an area of the projected flocs vs. their perimeter.

4.4.5 Fractal dimension in a relation to temperature

Surprisingly, a number of studies devoted to a participation of temperature through the fractal dimension parameter in the process of coagulation (specifically an impact on coagulation rate) are relatively rare. Out of them, the studies comparing coagulation kinetics at various temperatures are really scarce. It seems that the only exception is represented by a dairy industry (Horne 1999; Vetier et al. 1997, 2003). Using the fractal concept, Vetier et al. (1997, 2003) demonstrated how temperature and rate of acidification of casein micelles can affect aggregate formation and structure.

In the region of water treatment, Xiao et al. (2008, 2009) showed that low temperature slowed down the coagulation process as reflected in the decrease in aggregation rate and rate constants. Xiao et al. (2008) measured coagulation rate of kaolinite sus-

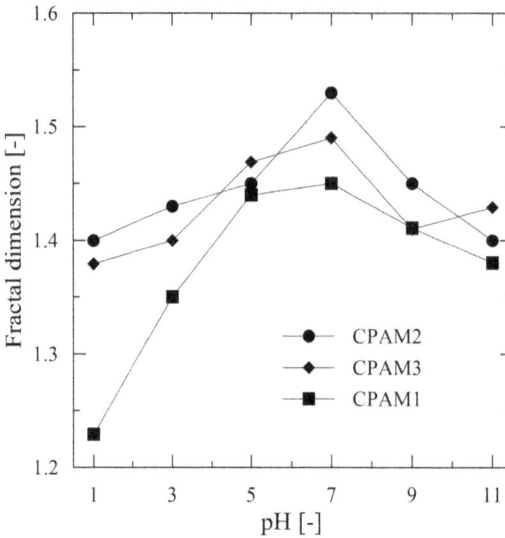

Fig. 4.50: A dependence of fractal dimension of clusters on pH for 3 coagulants poly(acrylamide-acryloyloxyethyl trimethyl ammonium chloride-butylacrylate) differing in (cationic degree [%], intrinsic viscosity [dL.d^{-1}]): CPAM1 (38, 1.7), CPAM2 (40, 2.38), and CPAM3 (21, 2.41) (data taken from Sun et al. 2015a, Fig. 8).

pension at 2, 5, and 8 °C and compared it with the results obtained at an ambient temperature 22 °C. An obvious reduction of the coagulation rate with temperature resulted in creation of smaller flocs. Xiao et al. 2009 showed that the perimeter-based fractal dimension D_{pf} (area $\propto$ (perimeter)$^{2/Dpf}$) increased from 1.19 to 1.33 as the temperature was downshifted from 22 to 2 °C. The aggregation rate was found to decline by 1–8% for every 1 °C from 22 °C to 2 °C. The resulting solution is to prolong a coagulation time (specifically in a winter period).

Using the fractal dimension D_2 of the aggregates relating the projected area A_s and the characteristic length of the aggregates l ($A_s \propto l^{D_2}$), Yu et al. (2013) investigated the aggregation kinetics of kaolin with and without presence of humic acids at low temperature. They showed that the fractal dimension D_2 was always lower in the presence of HA when the rate of aggregation at low temperature significantly increased and caused a decrease in the number of small flocs (exhibiting low settling ability).

4.4.6 Fractal dimension vs coagulants

A fractal dimension representing a rate of compactness of the formed flocs naturally strongly subjects to the applied coagulants. More compact flocs exhibit stronger tendency to settling, and in this sense efficiency of the individual coagulants can be ordered. Cheng et al. (2011) applied three different coagulants (aluminium sulphate, polyaluminium chloride, and cationic polyacrylamide) for treatment of the same raw water containing kaolin component. In their jar tests, they used a digital image technique for the determination of fractal dimensions of flocs under various shearing G. Specifically, they related a floc area to its perimeter through a power-law expression, where an exponent represents a 1D fractal dimension (see Fig. 4.51). It is not without interest that a ratio of efficiencies of two coagulants (polyaluminium chloride and cationic polyacrylamide) is practically the same in the studied range of shearing ($G <$ 200 s^{-1}), viz. a ratio of the fractal dimensions D_{PACl}/D_{cPAM} attains approximately the constant value 1.15.

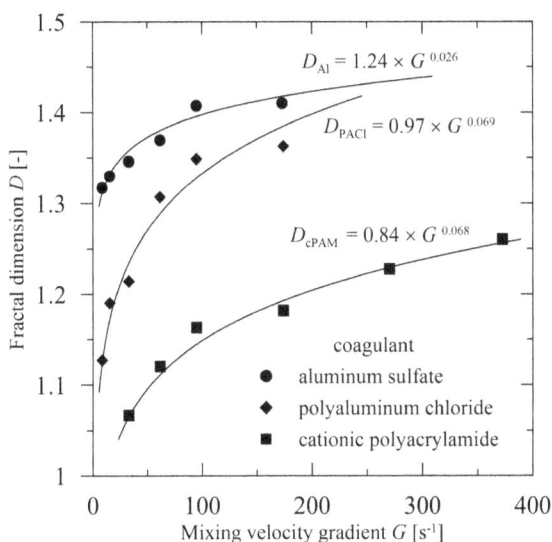

Within the figure:
$D_{Al} = 1.24 \times G^{0.026}$
$D_{PACl} = 0.97 \times G^{0.069}$
$D_{cPAM} = 0.84 \times G^{0.068}$

coagulant
● aluminum sulfate
◆ polyaluminum chloride
■ cationic polyacrylamide

Axis labels: Fractal dimension D [-]; Mixing velocity gradient G [s^{-1}]

Fig. 4.51: A dependence of fractal dimension on mixing velocity gradient for three coagulants applied to the identical raw water (data taken from Cheng et al. 2011, Fig. 11).

With a gradual increase of the mixing velocity gradient G, looser branches of flocs are removed. This reflects in more compactness of flocs, hence higher fractal dimension and strength. At the first stage (low G) relatively very loose flocs are created which results in easy cutting of loose branches and fast creation of more compact structures. It implies a higher increase of fractal dimension (see Fig. 4.51) in the range of lower G. With increasing shearing, a probability of branch cutting is progressively lower which corresponds to already moderate increase of fractal dimension. The limiting case (very high G) can be graphically depicted by an asymptote (a fractal dimension should tend to a constant value).

In this case it is necessary to abandon the preceding modelling using the power-law functions (see Fig. 4.51). Such modelling is useful only in a limited range of mixing velocity gradients ($G < 200$ s^{-1}). It fulfils the initial condition $D = 1$ at $G = 0$, however, completely fails for larger values of G non-predicting a finite limiting value of fractal dimension. The limiting values can be approximated by the following expression

$$D = 1 + a \times \tanh(G/b), \tag{4.86}$$

where the parameters a and b have a clear physical meaning. The parameter a participates in a limiting value (= 1 + a) of fractal dimension. The parameter b contributes to a value a/b representing a derivative (an initial slope) at $G = 0$. The term 1 fulfils the initial condition ($D = 1$ at $G = 0$) (see Fig. 4.52).

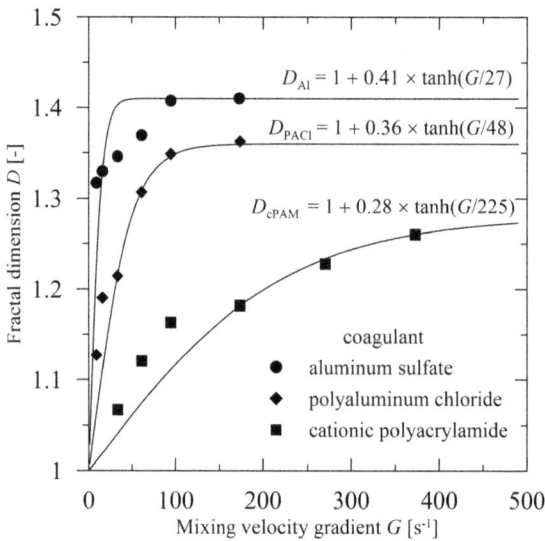

Fig. 4.52: Modelling of the limiting fractal dimensions for three coagulants (data taken from Cheng et al. 2011, Fig. 11).

4.4.7 Computer modelling of fractal aggregates

Alongside with the onset of fractal analysis and its application (usage) in the region of aggregate creation, various computer models were developed for a description of corresponding aggregate morphology characterising basic physical quantities (such as mass distribution, density, and porosity).

First contributions presenting computer simulations of fractal aggregates date back to the late 50s and early 60s (see Tab. 4.1). First, the clusters created from monodisperse spherical particles were enlarged by adding (joining) the individual particles, hence, the term particle-cluster aggregation models. According to a scheme of particle joining, three different basic mechanisms were consequently introduced: (1) reaction-limited, (2) ballistic, and (3) diffusion-limited. Each of these computer constructions of clusters significantly differs in morphology as documented by apparent differences in fractal values and therefore in expected physical quantities. In the following a brief description of the individual construction techniques will be outlined.

Tab. 4.1: Original references to six basic approaches to computer simulation of the process of aggregation.

Aggregation model	Reaction-limited		Ballistic		Diffusion-limited	
Particle-cluster	RLA	Eden (1956, 1961)	BA	Vold (1959a,b, 1960, 1963)	DLA	Witten and Sander (1981)
Cluster-cluster	RLCA	Meakin (1988b)	BCA	Sutherland (1967, 1970) Sutherland and Goodarz-Nia (1971)	DLCA	Finegold (1976) Kolb et al. (1983) Meakin (1983a)

Eden, M. (1956) A probabilistic model for morphogenesis. In: Symposium on Information Theory in Biology, Gatlinburg, Tennessee, 29–31 Oct, 1956, Eds. H.P. Yockey, R.L. Platzman, H. Quastler (Pergamon, New York, 1958), pp. 359–370.

Eden, M. (1961) A two-dimensional growth process. In: 4th Berkeley Symposium on Mathematical Statistics and Probability, 20 Jun – 30 Jul 1960, Ed. J. Neyman (University of California Press, Berkeley, 1961), pp. 223–239. Volume IV: Biology and the Problems of Health.

Finegold, L.X. (1976) Cell membrane fluidity: Modeling of particle aggregations seen in electron microscopy, Biochimica and Biophysica Acta 448, 393–398.

Kolb, M., Botet, R., Jullien, R. (1983) Scaling of kinetically growing clusters. Physical Review Letters 51, 1123–1126.

Meakin, P. (1983a) Formation of fractal clusters and networks by irreversible diffusion-limited aggregation. Physical Review Letters 51, 1119–1122.

Meakin, P. (1988b) Reaction-limited cluster-cluster aggregation in dimensionalities 2–10. Physical Reviews A 38, 4799–4814.

Sutherland, D.N. (1967) A theoretical model of floc structure, Journal of Colloid and Interface Science 25, 373–380.

Sutherland, D.N. (1970) Chain formation of fine particle aggregates. Nature 226, 1241–1242.

Sutherland, D.N., Goodarz-Nia, I. (1971) Floc simulation: the effect of collision sequence. Chemical Engineering Science 26, 2071–2085.

Vold, M. J. (1959a) Numerical approach to the problem of sediment volume. Journal of Colloid Science 14, 168–174.

Vold, M.J. (1959b) Sediment volume and structure in dispersions of anisometric particles. Journal of Physical Chemistry 63, 1608–1612.

Vold, M.J. (1960) The sediment volume in dilute dispersions of spherical particles. Journal of Physical Chemistry 64, 1616–1619.

Vold, M.J. (1963) Computer simulation of floc formation in a colloidal suspension. Journal of Colloid and Interface Science 18, 684–695.

Witten, T.A., Sander, L.M. (1981) Diffusion-limited aggregation, a kinetic critical phenomenon. Physical Review Letters 47, 1400–1403.

4.4.7.1 Reaction-limited particle-cluster aggregation model (RLA)

As the first step, an arbitrary cell in a blank square lattice is designed as occupied by particle. At the second step, one of the four adjacent cells is chosen as a site for another particle. Consequently, there are six neighbouring cells of two adjacent particles as the candidates for placing another particle. A growing cluster is enlarged by one particle at a time using an unoccupied cell along the unoccupied perimeter. The original Eden model took into consideration an occupancy of the neighbouring cells of potential new placing of newly added particle. There is also an alternative version that all candidates enter into the growth process with the same probability. Other random choices were also developed.

Typical morphology of a cluster generated by the RLA approach is depicted in Fig. 4.53. The cluster is very compact, tiny holes appearing along the outer perimeter are relatively fast filled. Compactness of the formed cluster results in a fractal dimension approaching the Euclidean one.

RLA

140 lattice units

Fig. 4.53: A 10,000 site cluster generated (RLA method) with equal growth probabilities of unoccupied sites (taken from Meakin 1988a, Fig. 5).

4.4.7.2 Ballistic particle-cluster aggregation model (BA)

A cluster is formed by monodispersed spherical particles successively added to a primary particle centred at the origin of the spherical coordinates (r, θ, φ) with a radius $r \geq 0$, polar angle θ $(0 \leq \theta \leq \pi)$, and azimuthal angle φ $(0 \leq \varphi < 2\pi)$. A location of next primary particle is based on a random choice of both angles θ and φ with a radial distance corresponding to a double (R) of a distance $(R/2)$ between the origin and – at that time – the most distant particle. A choice of random triple representing the Cartesian coordinates within the sphere R determines a point used for the construction of the straight line connecting this point with the last added primary particle at the distance R. The primary particle moves along this straight line (ballistic (linear) trajectory) towards the point. There are two possibilities: either the particle collides with a previous particle in the cluster (and irreversibly sticks) or passes untouched through the sphere (see Fig. 4.54).

Randomness of the choices of the angles θ and φ was questioned by Sutherland (1966) who argued by a sinusoidal character of the polar angle θ. In other words – taken from the viewpoint of spherical surface uniformity – the points at the 'polar'

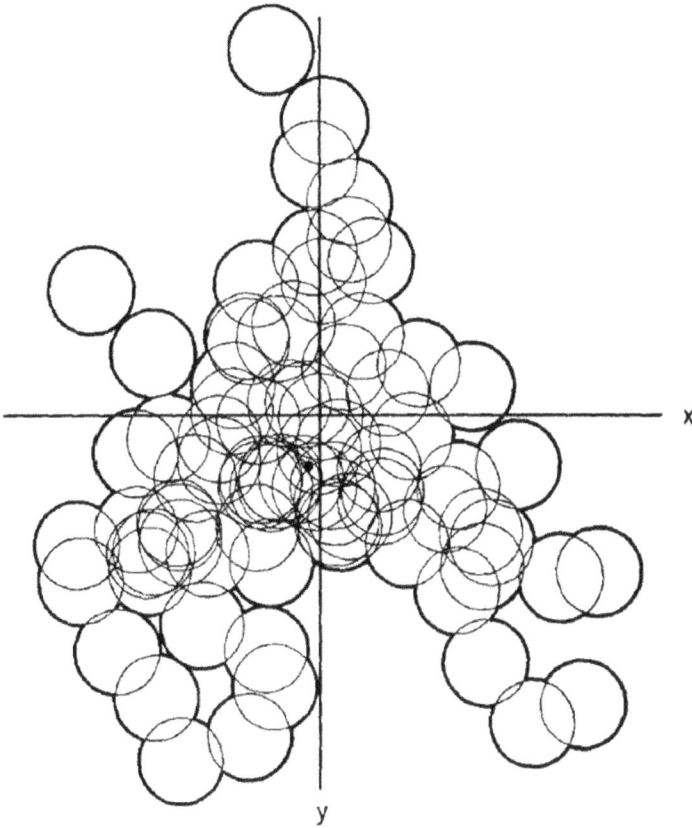

Fig. 4.54: Two-dimensional projection of primary particles forming a cluster using the BA model (taken from Vold 1963, Fig. 4).

regions (surface of spherical 'polar' caps) are preferred despite a substantially lower area of both caps in comparison with the central belt. Sutherland (1966) also introduced the corrections for levelling the randomness.

4.4.7.3 Diffusion-limited particle-cluster aggregation model (DLA)

In the first step a nucleation centre (a single spherical particle of diameter $2a$) is located at the origin of coordinate system. In the second step another particle is randomly located (in Fig. 4.55 on the circumference $r_{max} + 5$, where r_{max} is the maximum radius of a cluster). The particle randomly diffuses with the distance $2a$ and either approaches the cluster and sticks to it (trajectory t_1) at an unoccupied site on the perimeter or misses the cluster (trajectory t_2 is terminated) and another randomly placed particle, one at a time, is considered. The procedure is finished after reaching a required number of particles or a cluster size.

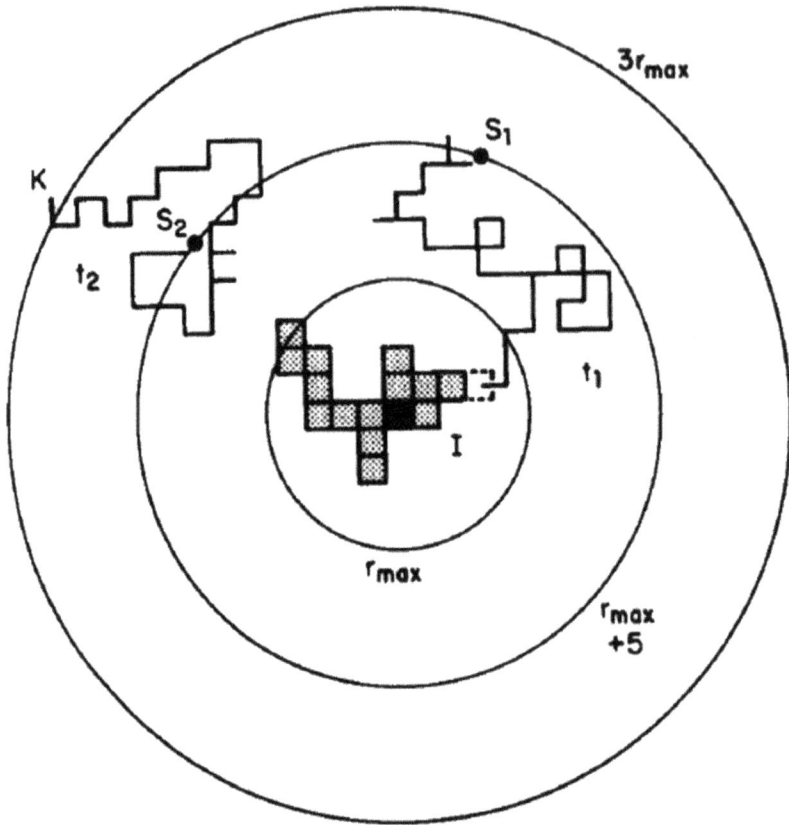

Fig. 4.55: An early stage in a square lattice model simulation of diffusion-limited aggregation. The nucleation centre in black is located at the coordinate origin, and last added particle is marked with a dashed square perimeter (taken from Meakin 1988a, Fig. 6).

Morphology of a cluster exhibits a dendritic design with relatively open architecture due to immediate sticking of the particles to a formed cluster (see Fig. 4.56). Usage of 10 colours documents an aggregation order, newly arrived particles have really a very limited possibility to penetrate inside a cluster and are located still further from a nucleation centre accumulated on the outside arms.

4.4.7.4 Summary of all three models (1)–(3)

Morphological generation of the clusters substantially differs which is naturally reflected in fractal characteristics (fractal dimensions in the ranges $1.7 \div 1.8$ and $2.2 \div 2.3$ for DLA and RLA, respectively). The RLA method creates in comparison with the DLA method much compact clusters with a minimum of open holes cumulated along the cluster periphery. The high level of compactness is proved by the density correlations, which are independent of distance in the limit of large size. Open dendritic structure

Fig. 4.56: A DLA cluster formed by 1,000,000 particles. Ten colours correspond to an aggregation order, each colour (starting from white) consequently represents 100,000 newly arrived particles (terminated by red) (taken from Sander 2000).

of DLA fractal objects is caused by fixed joining at first contact of an arriving particle with the cluster. This is well depicted in Fig. 4.57 (Meakin 1999), where each aggregate is formed by exactly 1,000 particles. Attention should be also paid to the extent of the individual aggregates indicated by a number of particles diameters.

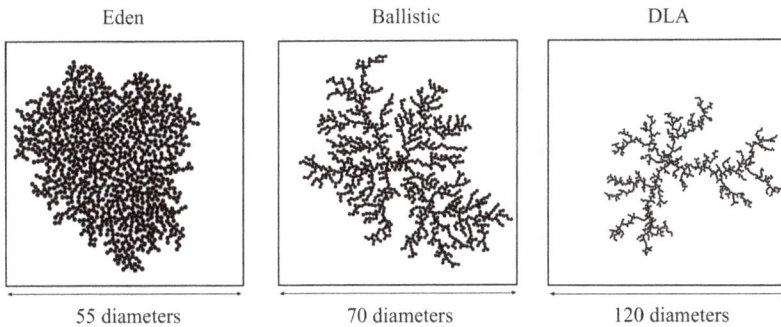

Eden	Ballistic	DLA
55 diameters	70 diameters	120 diameters

Fig. 4.57: Compactness vs. openness of the clusters generated through the RLA, ballistic, and DLA simulations, each method contains 1,000 particles (taken from Meakin 1999, Fig. 1).

The characteristics of all three clusters and their mutual comparison are presented in Meakin (1983b). An illustrative visual demonstration of the RLA and DLA methods using tennis balls is shown in Sabin et al. (2009).

Starting at the late 60s the contributions of various authors critically discussed the particle-cluster methods. Here are some points frequently analysed:

- adequateness of the models for a description of real situations (a distance from reality of the lattice models, inefficiency of a description of the growth of flocs in the suspension in the case of the ballistic models, very fragile connection between two sub-clusters through one particle only in the DLA model when exposed to the combined effects of gravity, Brownian movement, and fluid shear);
- in the case of larger clusters gradual adding of one particle seems to be inefficient in comparison with the adding of clusters;
- if the thermal motion is not competitive with the electrostatic repulsive barrier, particles repeatedly bounce each other before overcoming the electrostatic barrier when they immediately stick together; however, in this case the probability of sticking is reduced from the fixed 1 to an interval (0,1).

The method called the porous Eden model reflecting the preceding point was recently proposed by Guesnet et al. (2019). The additional capability of this model is to inactivate randomly any particle with a prescribed inactivation probability between 0 and 1. It enables deeper penetration of the particles into the formed cluster which can contribute to stiffer morphology; on the other hand it enables forming of porous structures. In combination, it strongly contributes to the efficiency of this method. The authors formulate five rules which have to be obeyed by the particles in forming the aggregate growth from the initial seed particle:

1. A particle is chosen randomly among active ones.
2. A growth direction is chosen randomly.
3. If there is enough space in this direction, a new particle is added in contact.
4. The chosen particle may be turned inactive with an inactivation probability p.
5. The process is iterated until a specified number of particles N is reached.

Figure 4.58 depicts four cluster morphologies for various values of probability (starting with the compact Eden model), where a number of particles attains 1,000 – the same as in Fig. 4.57.

The initial particle-cluster methods (RLA, ballistic, DLA) were consequently developed to the so-called cluster-cluster methods (for introducing literature see Tab. 4.1). Among other things it enabled simultaneous sticking of a newly added cluster at more than one particle. A detailed analysis of these methods is presented in Meakin and Jullien (1988), Meakin (1999), Stoll and Buffle (1995).

Originally all six computer methods were derived for monodisperse spherical particles. As in practice this situation is relatively seldom and an approximation replacing non-spherical (even irregular) particles by the spherical ones often fails, an intensive study of polydisperse systems and their computer simulation has been developed with the present advanced computational technique, see Morán et al. (2019), Tomchuk (2023), and references therein.

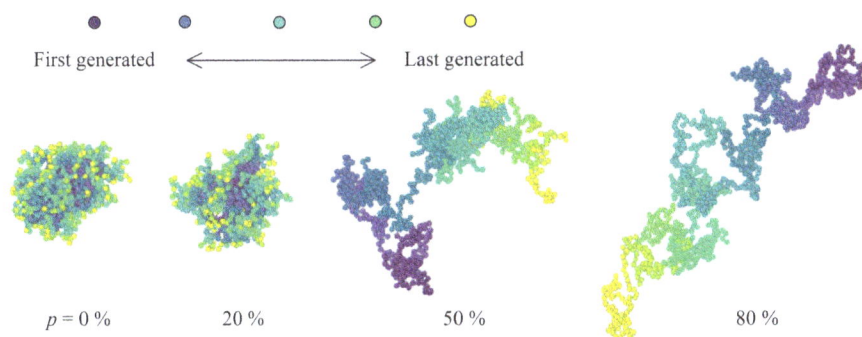

Fig. 4.58: An application of the porous Eden model. A tunable fractal dimension using a probability parameter p indicating a number of touches of a new particle before sticking to the formed cluster (taken from Guesnet et al. 2019, Fig. 1).

4.5 Summary of fractal approaches

A variety of experimental techniques applicable for a determination of the fractal dimension of aggregates is manifold. The attributes characterising the individual techniques can be expressed through the following points:

- Invasive vs. non-invasive technique. This point seems to be crucial as data processing of the experiments carried out with distorted suspended medium is of a limited value.
- Reproducibility of measurements. A large standard deviation of repeated measurements does not give a credit to the chosen measurements.
- Expenses. More expensive technique does not automatically imply better results in the specific measurements. Acquisition costs can be levelled by low maintenance expenses.
- A proper choice of an adequate experimental method. For instance, static light scattering supposes optically soft spherical monodisperse small particles with relatively low concentration. Using sedimentation, a fractal dimension can be assigned to large dense aggregates.
- A difference in applying the normal Cartesian or the log-log coordinates. The deviations in the log-log coordinates differ from those in the normal coordinates; it means that non-linear regression is transformed to the linearised regression with reduced physical interpretation.
- Carefulness whether to mix two spaces differing in dimensions (e.g. 3D Cartesian and generally non-integer fractal space) in deriving fractal dependencies. As an example, it is possible to introduce a passage from the expression relating aggregate mass and radius $M \propto r^{D_f}$ and a consequent substitution for a volume $V = (4/3)\pi r^3$.

In spite of cumulation of various incorrectness in determining a fractal dimension (non-conformity with the term 'fractal' introduced by Mandelbrot (1975), restrictive assumptions connected with the individual techniques not always in compliance with practice), a usefulness of a notion 'fractal dimension' for characterising aggregate morphology has proven its indisputable contribution.

Other question in evaluating fractal characterisation is in conformity of various techniques. Literature on this topic is really scarce. Tang et al. (2002) compared the results obtained with light scattering approach with those connected with the sedimentation process and found a discrepancy of fractal dimensions within very acceptable 12%. Jarvis et al. (2005) compared the same techniques, and in this case, they discuss the different results they obtained.

References

Antoniotti, L., Caldarola, F., Maiolo, M. (2020) Infinite numerical computing applied to Hilbert's, Peano's, and Moore's curves. Mediterranean Journal of Mathematics 17, 99. https://doi.org/10.1007/s00009-020-01531-5

Bache, D.H., Johnson, C., McGilligan, J.F., Rasool, E. (1997) A conceptual view of floc structure in the sweep floc domain. Water Science and Technology 36, 49–56. https://doi.org/10.1016/S0273-1223(97)00418-6

Beeckmans, J.M. (1964) The density of aggregated solid aerosol particles. Annals of Occupational Hygiene 7, 299–305. https://doi.org/10.1093/annhyg/7.3.299

Berry, M.V., Lewis, Z.V. (1980) On the Weierstrass-Mandelbrot fractal function. Proceedings of the Royal Society of London A370, 459–484. https://doi.org/10.1098/rspa.1980.0044

Biggs, S., Habgood, M., Jameson, G.J., Yan, Y. (2000) Aggregate structures formed via a bridging flocculation mechanism. Chemical Engineering Journal 80, 13–22. https://doi.org/10.1016/S1383-5866(00)00072-1

Boller, M., Blaser, S. (1998) Particles under stress. Water Science and Technology 37, 9–29. https://doi.org/10.2166/wst.1998.0367

Bremer, L.G.B., Bijsterbosch, B.H., Walstra, P., van Vliet, T. (1993) Formation, properties and fractal structure of particle gels. Advances in Colloid and Interface Science 46, 117–128. https://doi.org/10.1016/0001-8686(93)80037-C

Bubakova, P., Pivokonsky, M., Filip, P. (2013) Effect of shear rate on aggregate size and structure in the process of aggregation and at steady state. Powder Technology 235, 540–549. https://doi.org/10.1016/j.powtec.2012.11.014

Buczkowski, S., Kyriacos, S., Nekka, F., Cartilier, L. (1998) The modified box-counting method: Analysis of some characteristic parameters. Pattern Recognition 31, 411–418. https://doi.org/10.1016/S0031-3203(97)00054-X

Bushell, G.C., Yan, Y.D., Woodfield, D., Raper, J., Amal, R. (2002) On techniques for the measurement of the mass fractal dimension of aggregates. Advances in Colloid and Interface Science 95, 1–50. https://doi.org/10.1016/S0001-8686(00)00078-6

Buzoianu, M. (2000) Practical considerations on the traceability to conventional scales. Accreditation and Quality Assurance 5, 142–150. https://doi.org/10.1007/s007690050433

Chakraborti, R.K., Gardner, K.H., Atkinson, J.F., van Benschoten, J.E. (2003) Changes in fractal dimension during aggregation. Water Research 37, 873–883.

Chakraborti, R.K, Atkinson, J.F. (2020) Settling velocity analysis of natural suspended particles using fractal approach. Journal of Environmental Engineering 146, 04020138. https://doi.org/10.1061/(ASCE)EE.1943-7870.000182

Cheng, N.S. (2009) Comparison of formulas for drag coefficient and settling velocity of spherical particles. Powder Technology 189, 395–398. https://doi.org/10.1016/j.powtec.2008.07.006

Cheng, W.-P., Chen, P.H., Yu, R.-F., Hsieh, Y.-J., Huang, Y.W. (2011) Comparing floc strength using a turbidimeter. International Journal of Mineral Processing 100, 142–148. https://doi.org/10.1016/j.minpro.2011.05.010

Cheng, W.-P., Hsieh, Y.-J., Yu, R.-F., Huang, Y.-W., Wu, S.-Y., Chen, S.-M. (2010) Characterizing polyaluminum chloride (PACl) coagulation floc using an on-line continuous turbidity monitoring system. Journal of the Taiwan Institute of Chemical Engineers 41, 547–552. https://doi.org/10.1016/j.jtice.2010.01.002

Coulter, W.H. Means for counting particles suspended in a fluid. U.S. Patent 2,656,508, filed August 27, 1949 and issued October 20, 1953. https://patents.google.com/patent/US2656508A/en

de Arcangelis, L., Godanob, C., Grasso, J.R., Lippiello, E. (2016) Statistical physics approach to earthquake occurrence and forecasting. Physics Reports 628, 1–91. http://dx.doi.org/10.1016/j.physrep.2016.03.002

Dey, S, Ali, S.Z., Padhi E. (2019) Terminal fall velocity: the legacy of Stokes from the perspective of fluvial hydraulics. Proceedings of the Royal Society A 475, 20190277. https://doi.org/10.1098/rspa.2019.0277

Doty, P., Steiner, R.F. (1950) Light scattering and spectrophotometry of colloidal solutions. Journal of Chemical Physics 18, 1211–1220. https://doi.org/10.1063/1.1747913

Duan, Q., An, J., Mao, H., Liang, D., Li, H., Wang, S., Huang, C. (2021) Review about the application of fractal theory in the research of packaging materials. Materials 14, 860. https://doi.org/10.3390/ma14040860

Dyer, K.R., Manning, A.J. (1999) Observation of the size, settling velocity and effective density of flocs, and their fractal dimensions. Journal of Sea Research 41, 87–95. https://doi.org/10.1016/S1385-1101(98)00036-7

Eden, M. (1956) A probabilistic model for morphogenesis. In: Symposium on Information Theory in Biology, Gatlinburg, Tennessee, 29–31 Oct, 1956), Eds. H.P. Yockey, R.L. Platzman, H. Quastler (Pergamon, New York, 1958), pp. 359–370.

Eden, M. (1961) A two-dimensional growth process. In: 4th Berkeley Symposium on Mathematical Statistics and Probability, 20 Jun – 30 Jul 1960, Ed. J. Neyman (University of California Press, Berkeley, 1961), pp. 223–239. Volume IV: Biology and the Problems of Health.

Einstein, A. (1906) Eine neue Bestimmung der Moleküldimensionen. Annalen der Physik 324, 289–306. https://doi.org/10.1002/andp.19063240204

Einstein, A. (1911) Berichtigung zu meiner Arbeit: "Eine neue Bestimmung der Moleküldimensionen". Annalen der Physik 339, 591–592. https://doi.org/10.1002/andp.19113390313

Eisenlauer, J., Horn, D. (1985) Fibre-optic sensor technique for flocculant dose control in flowing suspensions. Colloids and Surfaces 14, 121–134. https://doi.org/10.1016/0166-6622(85)80046-9

Elliott, A.D. (2020) Confocal microscopy: Principles and modern practices. Current Protocols in Cytometry 92, e68. https://doi.org/10.1002/cpcy.68

Falconer, K. (1990) Fractal Geometry: Mathematical Foundations and Applications. Wiley, Chichester, 288 pp.

Fard, E.G. (1984) A cumulant LBM approach for large eddy simulation of dispersion microsystems. Ph.D. Thesis, Technische Universität Carolo-Wilhelmina zu Braunschweig, Germany. https://www.researchgate.net/publication/319153891_A_Cumulant_LBM_approach_for_Large_Eddy_Simulation_of_Dispersion_Microsystems

Farias, T.L., Koylu, U.O., Carvalho, M.G. (1996) Range of validity of the Rayleigh-Debye-Gans theory for optics of fractal aggregates. Applied Optics 35, 6560–6567. https://doi.org/10.1364/AO.35.006560

Ferreira, C., Cardona, J., Agimelen, O., Tachtatzis, C., Andonovic, I., Sefcik, J., Chen, Z.C. (2020) Quantification of particle size and concentration using in-line techniques and multivariate analysis. Powder Technology 376, 1–11. https://doi.org/10.1016/j.powtec.2020.08.015

Ferri, F., D'Angelo, A., Lee, M., Lotti, A., Pigazzini, M.C., Singh, K., Cerbino, R. (2011) Kinetics of colloidal fractal aggregation by differential dynamic microscopy. European Physical Journal Special Topics 199, 139–148. https://doi.org/10.1140/epjst/e2011-01509-9

Finegold, L.X. (1976) Cell membrane fluidity: Modeling of particle aggregations seen in electron microscopy, Biochimica and Biophysica Acta 448, 393–398. https://doi.org/10.1016/0005-2736(76)90252-2

Forrest, S.R., Witten, T.A., Jr. (1979) Long-range correlations in smoke-particle aggregates. Journal of Physics A: Mathematical and General 12, L109–L117. jav12i5pL109.pdf

Francois, S.B. (1987) Strength of aluminum hydroxide flocs. Water Research 21, 1023–1030. https://doi.org/10.1016/0043-1354(87)90023-6

Garcia-Aragon, J.A., Salinas-Tapia, H., Moreno-Guevara, J., Diaz-Palomarezi, V., Tejeda-Vega, S. (2014) Model for the settling velocity of flocs, application to an aquaculture recirculation tank. International Journal of Computational Methods and Experimental Measurements 2, 313–322. https://doi.org/10.2495/CMEM-V2-N3-313-322

Goldberger, A.L., West B.J. (1987) Fractals in physiology and medicine. The Yale Journal of Biology and Medicine 60, 421–435. https://www.ncbi.nlm.nih.gov/pmc/articles/PMC2590346/

Gorczyca, B., Ganczarczyk, J.J. (1996) Image analysis of alum coagulated mineral suspensions. Environmental Technology 17, 1361–1369. https://doi.org/10.1080/09593331708616505

Graham, M.D. (2003) The Coulter principle: Foundation of an industry. Journal of the Association for Laboratory Automation 8, 72–81. https://doi.org/10.1016/s1535-5535(03)00023-6 doi:10.1016/S1535-5535(03)00023-6

Greenwood, J., Rainey, T., Doherty, W.O.S. (2007) Light scattering study on the size and structure of calcium phosphate/hydroxyapatite flocs formed in sugar solutions. Journal of Colloid and Interface Science 306, 66–71. https://doi.org/10.1016/j.jcis.2006.10.010

Gregory J (1998) The role of floc density in solid–liquid separation. Filtration and Separation 35, 367–371. https://doi.org/10.1016/S0015-1882(97)87417-4

Gregory, J. (2009) Monitoring particle aggregation processes. Advances in Colloid and Interface Science 147–148, 109–123. https://doi.org/10.1016/j.cis.2008.09.003

Gregory, D., Carlson, K. (2003) Relationship of pH and floc formation kinetics to granular media filtration performance. Environmental Science and Technology 37, 1398–1403. https://doi.org/10.1021/es025801b

Gregory, J., Nelson, D.W. (1986) Monitoring of aggregates in flowing suspensions. Colloids Surf. A 18, 175–188. https://doi.org/10.1016/0166-6622(86)80312-2

Guesnet, E., Dendievel, R., Jauffrès, D., Martin, C.L., Yrieix, B. (2019) A growth model for the generation of particle aggregates with tunable fractal dimension. Physica A 513, 63–73. https://doi.org/10.1016/j.physa.2018.07.061

Halmos, P.R. (1974) Measure Theory. Springer, New York-Heidelberg-London.

Hasegawa, M., Liu, J., Okuda, K., Nunobiki, M. (1996) Calculation of the fractal dimensions of machined surface profiles. Wear 192, 40–45. https://doi.org/10.1016/0043-1648(95)06768-X

Heath, A.R., Fawell, P.D., Bahri, P.A., Swift, J.D. (2002) Estimating average particle size by Focused Beam Reflectance Measurement (FBRM). Particle & Particle Systems Characterization 19, 84–95. https://doi.org/10.1002/1521-4117(200205)19:2<84::AID-PPSC84>3.0.CO;2-1

Hongli, Y., Minqiang, F., Airong, L., Lianping, D. (2015) General formulas for drag coefficient and settling velocity of sphere based on theoretical law. International Journal of Mining Science and Technology 25, 219–223. https://doi.org/10.1016/j.ijmst.2015.02.009

Horne, D.S. (1987) Determination of the fractal dimension using turbidimetric techniques. Faraday Discussions of the Chemical Society 83, 259–270. https://doi.org/10.1039/DC9878300259

Horne, D.S. (1999) Formation and structure of acidified milk gels. International Dairy Journal 9, 261–268. https://doi.org/10.1016/S0958-6946(99)00072-2

Jarvis, P., Jefferson, B., Gregory, J., Parsons, S.A. (2005) A review of floc strength and breakage. Water Research 39, 3121–3137. https://doi.org/10.1016/j.watres.2005.05.022

Jarvis, P., Parsons, A., Henderson, R., Nixson, N., Jefferson, B. (2008) The practical application of fractal dimension in water treatment practice – the impact of polymer dosing. Separation Science and Technology 43, 1785–1797. https://doi.org/10.1080/01496390801974506

Jiang, Q., Logan, B.E. (1991) Fractal dimensions of aggregates determined from steady-state size distributions. Environmental Science and Technology 25, 2031–2038. https://doi.org/10.1021/es00024a007

Johnson, C.P., Li, X., Logfan, A. (1996) Settling velocities of fractal aggregates. Environmental Science and Technology 30, 1911–1918. https://doi.org/10.1021/es950604g

Jullien, R. (1992) From Guinier to fractals. Journal de Physique I 2, 759–770. https://doi.org/10.1051/jp1:1992178

Khelifa, A., Hill, P.S. (2006) Models for effective density and settling velocity of flocs. Journal of Hydraulic Research 44, 390–401. https://doi.org/10.1080/00221686.2006.9521690

Kilps, J.R., Logan, B.E., Alldredge, A.L. (1994) Fractal dimensions of marine snow determined from image analysis of in situ photographs. Deep Sea Research Part I: Oceanographic Research Papers 41, 1159–1169. https://doi.org/10.1016/0967-0637(94)90038-8

Kitchener, B.G.B., Wainwright, J., Parsons, A.J. (2017) A review of the principles of turbidity measurement. Progress in Physical Geography – Earth and Environment 41, 620–642. https://doi.org/10.1177/0309133317726540

Kolb, M., Botet, R., Jullien, R. (1983) Scaling of kinetically growing clusters, Physical Review Letters 51, 1123–1126. https://doi.org/10.1103/PhysRevLett.51.1123

Kovalsky, P., Bushell, G. (2005) In situ measurement of fractal dimension using focussed beam reflectance measurement. Chemical Engineering Journal 111, 181–188. https://doi.org/10.1016/j.cej.2005.02.020

Kramer, O.J.I., de Moel, P.J., Raaghav, S.K.R., Baars, E.T., van Vugt, W.H., Breugem, W.-P., Padding, J.T., van der Hoek, J.P. (2021) Can terminal settling velocity and drag of natural particles in water ever be predicted accurately? Drinking Water Engineering and Science 14, 53–71 + Supplement. https://doi.org/10.5194/dwes-14-53-2021

Kranenburg, C. (1994) The fractal structure of cohesive sediment aggregates. Estuarine Coastal Shelf Science 39, 1665–1680. doi: http://dx.doi.org/10.1016/S0272-7714(06)80002-8

Lau, Y.L., Krishnappan, B.G. (1994) Comparison of particle size measurements made with a water elutriation apparatus and a Malvern particle size analyzer. NWRI Contribution No. 94–82, Burlington, Ontario: Aquatic Ecosystem Protection Branch, National Water Research Institute, Canada.0020. https://publications.gc.ca/site/eng/9.875913/publication.html

Lee, C., Kramer, T.A. (2004) Prediction of three-dimensional fractal dimensions using the two-dimensional properties of fractal aggregates. Advances in Colloid and Interface Science 112, 49–57. https://doi.org/10.1016/j.cis.2004.07.001

Li, J., Du, Q., Sun, C. (2009) An improved box-counting method for image fractal dimension estimation. Pattern Recognition 42, 2460–2469. https://doi.org/10.1016/j.patcog.2009.03.001

Li, M., Wilkinson, D. (2005) Determination of non-spherical particle size distribution from chord length measurements. Part 1: Theoretical analysis. Chemical Engineering Science 60, 3251–3265. https://doi.org/10.1016/j.ces.2005.01.008

Li, X.-Y., Logan, B.E. (2000) Settling and coagulating behaviour of fractal aggregates. Water Science and Technology 42, 253–258. https://doi.org/10.2166/wst.2000.0388

Lin, WW; Sung, SS; Lee, DJ; Chen, YP; Chen, DS; Lee, SF (2003) Coagulation of humic-kaolin-PACI aggregates. Water Science and Technology 47, 145–152. https://doi.org/10.2166/wst.2003.0038

Liu, Y., Chen, L., Wang, H., Jiang, L., Zhang, Y., Zhao, J., Wang, D., Zhao, Y., Song, Y. (2014) An improved differential box-counting method to estimate fractal dimensions of gray-level images. Journal of Visual Communication and Image Representation 25, 1102–1111. https://doi.org/10.1016/j.jvcir.2014.03.008

Logan, B.E., Kilps, J.R. (1995) Fractal dimensions of aggregates formed in different fluid mechanical environments. Water Research 29, 443–453. https://doi.org/10.1016/0043-1354(94)00186-B

Lopez-Exposito, P., Negro, C., Blanco, A. (2019) Direct estimation of microalgal flocs fractal dimension through laser reflectance and machine learning. Algal Research 37, 240–247. https://doi.org/10.1016/j.algal.2018.12.007

Ma, Y., Shi, W., Peng, C.-K., Yang, A.C. (2018) Nonlinear dynamical analysis of sleep electroencephalography using fractal and entropy approaches. Sleep Medicine Reviews 37, 85–93. http://dx.doi.org/10.1016/j.smrv.2017.01.003

Maggi, F. (2007) Variable fractal dimension: A major control for floc structure and flocculation kinematics of suspended cohesive sediment. Journal of Geophysical Research 112, C07012. https://doi.org/10.1029/2006JC003951

Mandelbrot, B. (1967) How long is the coast of Britain? Statistical self-similarity and fractional dimension. Science 156 (3775), 636–638. https://www.science.org/doi/10.1126/science.156.3775.636

Mandelbrot, B. (1975) Les objets fractals: forme, hasard et dimension. Flammarion, Paris.

Mandelbrot, B.B. (1982) The Fractal Geometry of Nature. Freeman, New York, 1982.

Masliyah, J.H., Polikar, M. (1980) Terminal velocity of porous spheres. Canadian Journal of Chemical Engineering 58, 299–302. https://doi.org/10.1002/cjce.5450580303

Matos, T., Martins, M.S., Henriques, R., Goncalves, L.M. (2024) A review of methods and instruments to monitor turbidity and suspended sediment concentration. Journal of Water Process Engineering 64, 105624. https://doi.org/10.1016/j.jwpe.2024.105624

Meakin, P. (1983a) Formation of fractal clusters and networks by irreversible diffusion-limited aggregation. Physical Review Letters 51, 1119–1122. https://doi.org/10.1103/PhysRevLett.51.1119

Meakin, P. (1983b) The Vold-Sutherland and Eden models of cluster formation. Journal of Colloid and Interface Science 96, 415–424. https://doi.org/10.1016/0021-9797(83)90044-9

Meakin, P. (1988a) Fractal aggregates. Advances in Colloid and Interface Science 28, 249–331. https://doi.org/10.1016/0001-8686(87)80016-7

Meakin, P. (1988b) Reaction-limited cluster-cluster aggregation in dimensionalities 2–10. Physical Reviews A 38, 4799–4814. https://doi.org/10.1103/PhysRevA.38.4799

Meakin, P. (1991) Fractal aggregates in geophysics. Reviews of Geophysics 29, 317–354. https://doi.org/10.1029/91RG00688

Meakin, P. (1999) A historical introduction to computer models for fractal aggregates. Journal of Sol-Gel Science and Technology 15, 97–117. https://doi.org/10.1023/A:1008731904082

Meakin, P.; Jullien, R. (1988) The effects of restructuring on the geometry of clusters formed by diffusion-limited, ballistic, and reaction-limited cluster-cluster aggregation. Journal of Chemical Physics 89, 246–250. https://doi.org/10.1063/1.455517

Merkus, H.G.: Particle Size Measurements: Fundamentals, Practice, Quality. Springer, London, 2009.

Minsky, M. (1957) Microscopy Apparatus. U.S. Patent no. 3013467 (awarded 19 Dec, 1961). https://patents.google.com/patent/US3013467A/en

Moghaddam, R.F., Cheriet, M. (2015) Modified Hausdorff fractal dimension (MHFD). arXiv:1505.03493. https://doi.org/10.48550/arXiv.1505.03493

Morán, J., Fuentes, A., Liu, F., Yon, J. (2019) FracVAL: An improved tunable algorithm of cluster–cluster aggregation for generation of fractal structures formed by polydisperse primary particles. Computer Physics Communications 239, 225–237. https://doi.org/10.1016/j.cpc.2019.01.015

Moruzzi, R.B., Bridgeman, J., Silva, P.A.G. (2020a) A combined experimental and numerical approach to the assessment of floc settling velocity using fractal geometry. Water Science and Technology 81, 915–924. https://doi.org/10.2166/wst.2020.171

Moruzzi, R.B., Campos, L.C., Sharific, S., da Silva, P.G., Gregory, J. (2020b) Nonintrusive investigation of large Al-kaolin fractal aggregates with slow settling velocities. Water Research 185, 116287. https://doi.org/10.1016/j.watres.2020.116287

Moruzzi, R.B., de Oliveira, A.L., da Conceição, F.T., Gregory, J., Campos, L.C. (2017) Fractal dimension of large aggregates under different flocculation conditions. Science of the Total Environment 609, 807–814. https://doi.org/10.1016/j.scitotenv.2017.07.194

Naceradska, J., Pivokonska, L., Pivokonsky, M. (2019) On the importance of pH value in coagulation. Journal of Water Supply: Research and Technology-Aqua 68, 222–230. https://doi.org/10.2166/aqua.2019.155

Nayak, S.R., Mishra, J., Palai, G. (2019) Analysing roughness of surface through fractal dimension: A review. Image and Vision Computing 89, 21–34. https://doi.org/10.1016/j.imavis.2019.06.015

Neale, G., Epstein, N., Nader, W. (1973) Creeping flow relative to permeable spheres. Chemical Engineering Science 28, 1865–1874. https://doi.org/10.1016/0009-2509(73)85070-5

Omar, A.F., MatJafri, M.Z. (2009) Turbidimeter design and analysis: A review on optical fiber. Sensors 9, 8311–8335. https://doi.org/10.3390/s91008311

Paddock, S.W. (2000) Principles and practices of laser scanning confocal microscopy. Molecular Biotechnology 16, 127–149. https://doi.org/10.1385/MB:16:2:127

Panigrahy, C., Seal, A., Mahato, N.K., Bhattacharjee, D. (2019) Differential box counting methods for estimating fractal dimension of gray-scale images: A survey. Chaos, Solitons and Fractals 126, 178–202. https://doi.org/10.1016/j.chaos.2019.06.007

Parker, D.S., Asce, A.M., Kaufman, W.J., Jenkins, D. (1972) Floc breakup in turbulent flocculation processes. Journal of the Sanitary Engineering Division, Proc. ASCE 98, 79–99. https://doi.org/10.1061/JSEDAI.0001389

Peano, G. (1890) Sur une courbe, qui remplit toute une aire plane. Mathematische Annalen 36, 157–160. https://doi.org/10.1007/BF01199438

Pivokonsky, M., Novotna, K., Petricek, R., Cermakova, L., Prokopova, M., Naceradska, J. (2024) Fundamental chemical aspects of coagulation in drinking water treatment – Back to basics. Journal of Water Process Engineering 57, 104660. https://doi.org/10.1016/j.jwpe.2023.104660

Pope, R.M., Fry, E.S. (1997) Absorption spectrum (380–700 nm) of pure water. II. Integrating cavity measurements. Applied Optics 36, 8710–8723. https://opg.optica.org/ao/fulltext.cfm?uri=ao-36-33-8710&id=63107

Qin, J.-J., Oo, M.H., Kekre, K.A., Knops, F., Miller, P. (2006) Impact of coagulation pH on enhanced removal of natural organic matter in treatment of reservoir water. Separation and Purification Technology 49, 295–298. https://doi.org/10.1016/j.seppur.2005.09.016

Ren, B., Weitzel, K.A., Duan, X., Nadagouda, M.N., Dionysiou, D.D. (2022) A comprehensive review on algae removal and control by coagulation-based processes: mechanism, material, and application. Separation and Purification Technology 293, 121106. https://doi.org/10.1016/j.seppur.2022.121106

Ruf, A., Worlitschek, J., Mazzotti, M. (2000) Modeling and experimental analysis of PSD measurements through FBRM. Particle & Particle Systems Characterization 17, 167–179. https://doi.org/10.1002/1521-4117(200012)17:4<167::AID-PPSC167>3.0.CO;2-T

Sabin, J., Bandin, M., Prieto, G., Sarmiento, F. (2009) Fractal aggregates in tennis ball systems. Physics Education 44, 499–502. https://iopscience.iop.org/article/10.1088/0031-9120/44/5/008

Sadar, M.J. (1998) Turbidity science. Technical Information Series-Booklet No. 11, Hach Company, Colorado, USA, 26 pp. https://www.hach.com/asset-get.download-en.jsa?code=61792

Sander, L.M. (2000) Diffusion-limited aggregation: A kinetic critical phenomenon? Contemporary Physics 41, 203–218. https://doi.org/10.1080/001075100409698

Schaefer, D.W., Martin, J.E., Wiltzius, P., Cannell, D.S. (1984) Fractal geometry of colloidal aggregates. Physical Review Letters 52, 2371–2374. https://doi.org/10.1103/PhysRevLett.52.2371

Schiller, L., Naumann, A. (1933) Über die grundlegenden Berechnungen bei der Schwerkraftaufbereitung. (Fundamental calculations in gravitational processing) Zeitschrift des Vereines deutscher Ingenieure 77, 318–320. https://api.semanticscholar.org/CorpusID:222424630

Senesi, N., Rizzi, F.R., Dellino, P., Acquafredda, P. (1996) Fractal dimension of humic acíds in aqueous suspension as a function of pH and time. Soil Science Society of America Journal 60, 1773–1780. https://doi.org/10.2136/sssaj1996.03615995006000060023x

Senesi, N., Rizzi, F.R., Dellino, P., Acquafredda, P. (1997) Fractal humic acids in aqueous suspensions at various concentrations, ionic strengths, and pH values. Colloids and Surfaces A: Physicochemical and Engineering Aspects 127, 57–68. https://doi.org/10.1016/S0927-7757(96)03949-0

Sillanpaa, M., Chaker Ncibi, M., Matilainen, A., Vepsalainen, M. (2018) Removal of natural organic matter in drinking water treatment by coagulation: A comprehensive review. Chemosphere 190, 54–71. https://doi.org/10.1016/j.chemosphere.2017.09.113

Sokołowska, Z., Sokołowski, S. (1999) Influence of humic acid on surface fractal dimension of kaolin: analysis of mercury porosimetry and water vapour adsorption data. Geoderma 88, 233–249. https://doi.org/10.1016/S0016-7061(98)00107-4

Sorensen, C.M. (2001) Light scattering by fractal aggregates: A review. Aerosol Science and Technology 35, 648–687. https://doi.org/10.1080/02786820117868

Sorensen, C.M., Lu, N., Cai, J. (1995) Fractal cluster size distribution measurement using static light scattering. Journal of Colloid and Interface Science 174, 456–460. https://doi.org/10.1006/jcis.1995.1413

Spicer, P.T., Keller, W., Pratsinis, S.E. (1996) The effect of impeller type on floc size and structure during shear-induced flocculation. Journal of Colloid and Interface Science 184, 112–122. https://doi.org/10.1006/jcis.1996.0601

Spicer, P.T., Pratsinis, S.E., Raper, J., Amal, R., Bushell, G., Meesters, G. (1998) Effect of shear schedule on particle size, density, and structure during flocculation in stirred tanks. Powder Technology 97, 26–34. https://doi.org/10.1016/S0032-5910(97)03389-5

Stoll, S., Buffle, J. (1995) Computer simulations of colloids and macromolecules aggregate formation. Chimia 49, 300–307. https://doi.org/10.2533/chimia.1995.300

Stone, S., Bushell, G., Amal, R., Ma, Z., Merkus, H.G., Scarlett, B. (2002) Characterization of large fractal aggregates by small-angle light scattering. Measurement Science and Technology 13, 357–364. https://doi.org/10.1088/0957-0233/13/3/318

Sun, Y., Fan, W., Zheng, H., Zhang, Y., Li, F., Chen, W. (2015a) Evaluation of dewatering performance and fractal characteristics of alum sludge. PLoS ONE 10, e0130683. https://doi.org/10.1371/journal.pone.0130683

Sun, Y.J., Zheng, H.L., Xiong, Z.P., Wang, Y.L., Tang, X.M., Chen, W., Ding, Y. (2015b) Algae removal from raw water by flocculation and the fractal characteristics of flocs. Desalination and Water Treatment 56, 894–904. https://doi.org/10.1080/19443994.2014.944586

Sutherland, D.N. (1966) Letter to the Editors: Comments on Vold's simulation of floc formation. Journal of Colloid and Interface Science 22, 300–302. https://doi.org/10.1016/0021-9797(66)90037-3

Sutherland, D.N. (1967) A theoretical model of floc structure. Journal of Colloid and Interface Science 25, 373–380. https://doi.org/10.1016/0021-9797(67)90043-4

Sutherland, D.N. (1970) Chain formation of fine particle aggregates. Nature 226, 1241–1242. https://doi.org/10.1038/2261241a0

Sutherland, D.N., Goodarz-Nia, I. (1971) Floc simulation: The effect of collision sequence. Chemical Engineering Science 26, 2071–2085. https://doi.org/10.1016/0009-2509(71)80045-3

Tang, P., Greenwood, J., Raper, J.A. (2002) A model to describe the settling behavior of fractal aggregates. Journal of Colloid and Interface Science 247, 210–219. https://doi.org/10.1006/jcis.2001.8028

Tang, P., Raper, J.A. (2002) Modelling the settling behaviour of fractal aggregates – a review. Powder Technology 123, 114–125. https://doi.org/10.1016/S0032-5910(01)00448-X

Tata, B.V.R., Raj, B. (1998) Confocal laser scanning microscopy: Applications in material science and technology. Bulletin of Materials Science 21, 263–278. https://doi.org/10.1007/BF02744951

Tence, M., Chevalier, J.P., Jullien, R. (1986) On the measurement of the fractal dimension of aggregated particles by electron microscopy: experimental method, corrections and comparison with numerical models. Journal de Physique 47, 1989–1998. https://hal.science/jpa-00210394

Teng, X., Li, F., Lu, C. (2020) Visualization of materials using the confocal laser scanning microscopy technique. Chemical Society Reviews 49, 2408–2425. https://doi.org/10.1039/C8CS00061A

Thill, A., Veerapaneni, S., Simon, B., Wiesner, M., Bottero, J.Y., Snidaro, D. (1998) Determination of structure of aggregates by confocal scanning laser microscopy. Journal of Colloid and Interface Science 204, 357–362. https://doi.org/10.1006/jcis.1998.5570

Tomchuk, O. (2023) Models for simulation of fractal-like particle clusters with prescribed fractal dimension. Fractal and Fractional 7, 866. https://doi.org/10.3390/fractalfract7120866

Tzoganakis, C., Price, B.C., Hatzikiriakos, S.G. (1993) Fractal analysis of the sharkskin phenomenon in polymer melt extrusion. Journal of Rheology 37, 355–366. https://doi.org/10.1122/1.550447

Vahedi, A., Gorczyca, B. (2011) Application of fractal dimensions to study the structure of flocs formed in lime softening process. Water Research 45, 545–556.

Vahedi, A., Gorczyca, B. (2012) Predicting the settling velocity of flocs formed in water treatment using multiple fractal dimensions. Water Research 46, 4188–4194. https://doi.org/10.1016/j.watres.2012.04.031

Vahedi, A., Gorczyca, B. (2014) Settling velocities of multifractal flocs formed in chemical coagulation process. Water Research 53, 322–328. https://doi.org/10.1016/j.watres.2014.01.008

Vetier, N., Banon, S., Chardot, V., Hardy, J. (2003) Effect of temperature and aggregation rate on the fractal dimension of renneted casein aggregates. Journal of Dairy Science 86, 2504–2507. https://doi.org/10.3168/jds.S0022-0302(03)73844-2

Vetier, N., Desobry-Banon, S., Ould Eleya, M.M., Hardy, J. (1997) Effect of temperature and acidification rate on the fractal dimension of acidified casein aggregates. Journal of Dairy Science 80, 3161–3166. https://doi.org/10.3168/jds.S0022-0302(97)76287-8

Vold, M. J. (1959a) Numerical approach to the problem of sediment volume. Journal of Colloid Science 14, 168–174. https://doi.org/10.1016/0095-8522(59)90041-8

Vold, M.J. (1959b) Sediment volume and structure in dispersions of anisometric particles. Journal of Physical Chemistry 63, 1608–1612. https://doi.org/10.1021/j150580a011

Vold, M.J. (1960) The sediment volume in dilute dispersions of spherical particles. Journal of Physical Chemistry 64, 1616–1619. https://doi.org/10.1021/j100840a004

Vold, M.J. (1963) Computer simulation of floc formation in a colloidal suspension. Journal of Colloid and Interface Science 18, 684–695. https://doi.org/10.1016/0095-8522(63)90061-8

Waite, T. (1999) Measurement and implications of floc structure in water and wastewater treatment. Colloids and Surfaces A: Physicochemical and Engineering Aspects 151, 27–41. https://doi.org/10.1016/S0927-7757(98)00634-7

Wang, D., Wu, R., Jiang, Y., Chow, C.W.K. (2011a) Characterization of floc structure and strength: role of changing shear rates under various coagulation mechanisms. Colloids and Surfaces A: Physicochemical and Engineering Aspects 379, 36–42. https://doi.org/10.1016/j.colsurfa.2010.11.048

Wang, L., Zeng, X., Yang, H., Lv, X., Guo, F., Shi, Y., Hanif, A. (2021) Investigation and application of fractal theory in cement-based materials: A review. Fractal and Fractional 5, 247. https://doi.org/10.3390/fractalfract5040247

Wang, Y.L., Feng, J., Dentel, S.K., Lu, J., Shi, B.Y., Wang, D.S. (2011b) Effect of polyferric chloride (PFC) doses and pH on the fractal characteristics of PFC–HA flocs. Colloids and Surfaces A: Physicochemical and Engineering Aspects 379, 51–61. https://doi.org/10.1016/j.colsurfa.2010.11.044

Winslow, D.N. (1985) The fractal nature of the surface of cement paste. Cement and Concrete Research 15, 817–824. https://doi.org/10.1016/0008-8846(85)90148-6

Witten, T.A., Sander, L.M. (1981) Diffusion-limited aggregation, a kinetic critical phenomenon. Physical Review Letters 47, 1400–1403. https://doi.org/10.1103/PhysRevLett.47.1400

Wright, S.J., Centonze, V.E., Stricker, S.A., DeVries, P.J., Paddock, S.W., Schatten, G. (1993) Introduction to Confocal Microscopy and Three-Dimensional Reconstruction. Methods in Cell Biology, Vol. 38: Cell Biological Applications of Confocal Microscopy (Ed. B. Matsumoto), Academic Press, New York 1993, Chapter 1, pp. 1–45.

Wu, J., Jin, X., Mi, S., Tang, J. (2020) An effective method to compute the box-counting dimension based on the mathematical definition and intervals. Results in Engineering 6, 100106. https://doi.org/10.1016/j.rineng.2020.100106

Wu, M., Wang, W., Shi, D., Song, Z., Li, M., Luo, Y. (2021) Improved box-counting methods to directly estimate the fractal dimension of a rough surface. Measurement 177, 109303. https://doi.org/10.1016/j.measurement.2021.109303

Xiao, F., Howard Huang, J.-C., Zhang, B.-J., Cui, C.-W. (2009) Effects of low temperature on coagulation kinetics and floc surface morphology using alum. Desalination 237, 201–213. https://doi.org/10.1016/j.desal.2007.12.033

Xiao, F., Ma, J., Yi, P., Huang, J.C.H. (2008) Effects of low temperature on coagulation of kaolinite suspensions. Water Research 42, 2983–2992. https://doi.org/10.1016/j.watres.2008.04.013

Xu, H., Jiao, R., Xiao, F., Wang, D. (2014) Effects of different coagulants in treatment of TiO2–humic acid (HA) water and the aggregate characterization in different coagulation conditions. Colloids and Surfaces A: Physicochemical and Engineering Aspects 446, 213–223.

Yang, Z., Li, H., Yan, H., Wu, H., Yang, H., Wu, Q., Li, H., Li, A., Cheng, R. (2014) Evaluation of a novel chitosan-based flocculant with high flocculation performance, low toxicity and good floc properties. Journal of Hazardous Materials 276, 480–488. http://dx.doi.org/10.1016/j.jhazmat.2014.05.061

Yu, W., Erickson, K. (2008) Chord length characterization using focused beam reflectance measurement probe-methodologies and pitfalls. Powder Technology 185, 24–30. https://doi.org/10.1016/j.powtec.2007.09.011

Yu, W.-Z., Gregory, J., Li, G.-B., Qu, J.-H. (2013) Effect of humic acid on coagulation performance during aggregation at low temperature. Chemical Engineering Journal 223, 412–417. https://doi.org/10.1016/j.cej.2013.03.008

Yu, W.-Z., Gregory, J., Liu, H.-J., Qu, J.-H. (2014) Investigation of the property of kaolin–alum flocs at acidic pH. Colloids and Surfaces A: Physicochemical and Engineering Aspects 443, 177–181. https://doi.org/10.1016/j.colsurfa.2013.11.005

Yukselen, M.A., Gregory, J. (2002) Breakage and re-formation of alum flocs. Environmental Engineering Science 19, 229–236. https://doi.org/10.1089/109287502760271544

Zhang, H., Yang, L., Zang, X., Cheng, S., Zhang, X. (2019) Effect of shear rate on floc characteristics and concentration factors for the harvesting of Chlorella vulgaris using coagulation-flocculation sedimentation. Science of the Total Environment 688, 811–817. https://doi.org/10.1016/j.scitotenv.2019.06.321

Zhao, P., Ge, S., Chen, Z., Li, X. (2013) Study on pore characteristics of flocs and sludge dewaterability based on fractal methods (pore characteristics of flocs and sludge dewatering). Applied Thermal Engineering 58, 217–223. https://doi.org/10.1016/j.applthermaleng.2013.03.018

Zhao, Y.X., Gao, B.Y., Zhang, G.Z., Qi, Q.B., Wang, Y., Phuntsho, S., Kim, J.H., Shon, H.K., Yue, Q.Y., Li, Q. (2014) Coagulation and sludge recovery using titanium tetrachloride as coagulant for real water treatment: a comparison against traditional aluminum and iron salts. Separation and Purification Technology 130, 19–27. https://doi.org/10.1016/j.seppur.2014.04.015

Zheng, H., Zhu, G., Jiang, S., Tshukudu, T., Xiang, X., Zhang, P., He, Q. (2011) Investigations of coagulation–flocculation process by performance optimization, model prediction and fractal structure of flocs. Desalination 269, 148–156. https://doi.org/10.1016/j.desal.2010.10.054

List of abbreviations

AB	Acid-base
AOM	Algal organic matter
BA	Ballistic particle-cluster aggregation
CCC	Critical coagulation concentration
CCD	Central composite design or charge-coupled device
CLD	Chord length distribution
COM	Cellular organic matter
CPAM	Cationic polyacrylamide
CSLM	Confocal Scanning Laser Microscopy
DBP	Disinfection by-product
DLA	Diffusion-limited particle-cluster aggregation
DLVO	Derjaguin-Landau-Verwey-Overbeek
DMN	Dimethylnitrosamine
DOC	Dissolved organic carbon
DoE	Design of experiments
DWTP	Drinking water treatment plant
EDL	Electrical double layer
EfOM	Effluent organic matter
EOM	Extracellular organic matter
EPA	Environmental Protection Agency
FBRM	Focused beam reflectance measurement
FI	Flocculation Index
FNU	Formazin Nephelometric Unit
GAC	Granular activated carbon
HA	Humic acid
HAA	Haloacetic acid
HAA5	Five most frequent haloacetic acids
HAA9	Nine most frequent haloacetic acids
HOMO	Highest occupied molecular orbital
ISO	International Organization for Standardization
IUPAC	International Union of Pure and Applied Chemistry
JPG (JPEG)	Joint Photographic Experts Group
LUMO	Lowest occupied molecular orbital
MWT	Multi-wavelength turbidimetry
NDMA	N-nitrosodimethylamine
NOM	Natural organic matter
NTU	Nephelometric Transmission Unit
OFAT	One factor at a time
PAC	Powdered activated carbon or polyaluminum chloride
PACl	Polyaluminum chloride
PACS	Polyaluminum chloro-sulfate
PAS	Polyaluminum sulfate
PFC	Polyferric chloride
PFS	Polyferric sulfate
pH	Potential of hydrogen
Re	Reynolds number
RGB	Red-green-blue

https://doi.org/10.1515/9783111246765-005

RGD	Rayleigh-Gans-Debye
RLA	Reaction-limited particle-cluster aggregation
RSM	Response surface method
THM	Trihalomethane
TTHM	Total trihalomethane
vdW	van der Waals
WTP	Water treatment plant
XDLVO	eXtended DLVO

Index

adsorption 93
aggregate 43, 68, 127, 131, 138, 162, 174
approximation
– Debye-Hückel 24, 51
– Malmberg-Maryott 48
artificial intelligence 73

Born repulsion 31
Brownian motion 2, 47

Cantor set 119
carbon
– dissolved organic 85, 95
– granular activated 93
– powdered activated 93
chloride
– polyaluminum 69, 170
– ferric 94, 170
central composite design 100
chloramination 84
chlor(am)ination 106
chlorination 84, 108
chord length distribution 152, 154
coagulant dosage 70, 91, 170
coagulant 68, 93, 169, 172
coagulation 1, 43, 47, 60, 68, 84, 93, 124, 165, 171
coastline paradox 117
colloidal particle 9, 36, 43
complexation 93
confocal scanning laser microscopy 148
constant
– equilibrium 89
– Hamaker 11, 18
critical coagulation concentration 64
curve
– Koch 120
– Peano 122

Debye-Hückel approximation 24
Debye length 26, 51
depletion 36
destabilisation 43, 70, 88, 91
diffusion coefficient 2
diffusion equation 2
digital image analysis 144
disinfection by-product 84, 106
displacement of colloidal particles 4

Dorn effect 59
drag coefficient 59, 165

electrokinetic phenomena 49
electroosmosis 50
electrophoresis 50, 59
energy
– repulsive potential 22
– van der Waals attraction 11
energy barrier 28, 59
equation
– diffusion 2
– Navier-Stokes 52
– Poisson 53
– Poisson-Boltzmann 66
– Stokes-Einstein-Sutherland 3

factor
– form 161
– interparticle structure 161
– recovery 131, 158
– strength 131, 158
ferrate(VI) 106
flocculant 68, 84, 91, 108
flocculation 1, 6, 44, 68, 54, 93, 110, 165
flow
– electroosmotic 50
– electrophoretic 55
force
– Born 9, 31
– Debye 10
– Keesom 10, 20, 28
– London 10, 20, 28
– van der Waals 10
fractal 116
fractal dimension 116

Guinier regime 163
Guldberg-Waage law 89

haloacetic acid 106
hydrolysis 88
hydrophobic attraction 32
hyperboloid mixer 109

index
– flocculation 157

https://doi.org/10.1515/9783111246765-006

– Wentworth flatness 6
inner Helmholtz plane 24, 28, 44
interaction
– Lewis acid-base 29
– steric 34

jar test 70, 75, 94, 100, 108

layer
– electrical double 24, 45
– Guoy-Chapman 24, 44
– Helmholtz 24, 28
– Stern 24, 44
Lifschitz approach 19
light scattering transmission 154

matter
– algal organic 105
– cellular organic 105
– extracellular organic 105
– natural organic 87
measure
– Hausdorff 117, 123
– Lebesgue 117
Menger sponge 121
method
– box counting 138
– electrical 143
– focused beam reflectance 151
– one factor at a time 100
– optical 142, 144
– response surface 100
– sand box 142
mobility
– electroosmotic 53
– electrophoretic 55, 59
model
– ballistic 176
– DLA 177
– Eden 176, 180
– RLA 176

N-nitrosodimethylamine 107
network
– shallow neural 74
– deep neural 75
number
– Dukhin 61
– pH 84

– power 110
– Reynolds 3, 51, 165

outer Helmholtz plane 44, 61
ozone 106

parameter
– polymer-solvent interaction 34
– Hildebrand solubility 34
particle
– colloidal 4, 9
– non-spherical 5, 71, 127, 154
– spherical 5, 11, 125, 162
permittivity 20, 26, 48
plane
– inner Helmholtz 24, 28, 44, 51
– outer Helmholtz 24, 44, 61
polyaluminium chloride 69, 94, 170
polymer bridging 35
potassium permanganate 106
potential
– Lennard-Jones 22
– sedimentation 50, 58
– streaming 50, 56, 63
– zeta 45, 53, 56, 61
power 109
pre-oxidation 106

radius of gyration 126, 163
regression analysis 76, 103
relation
– Helmholtz-Smoluchowski 53
– Henry 54
– Hückel-Onsager 53
– Vogel-Fulcher-Tammann 4
rule
– Schulze-Hardy 44. 63, 65, 88
– inverse Schulze-Hardy 67

scattered intensity 159
self-similarity 119
shearing 43, 108, 111, 127, 143, 173
Sierpiński carpet 135
Sierpiński triangle 121
solvation 32
stability behaviour 47
static light scattering 159, 181
sulfate
– aluminium 91, 94, 169

– polyferric 69
sweep coagulation 91

theory
– DLVO 27
– Rayleigh-Gans-Debye 160
– XDLVO 28
trihalomethane 106
turbidity 72, 76, 91, 101, 154, 170

van der Waals attraction energy 11
velocity
– gradient 110, 173
– sedimentation 7, 72, 129
– settling 7, 125, 164
viscosity 40, 47, 51, 62, 125, 165

water density distribution function 45

www.ingramcontent.com/pod-product-compliance
Lightning Source LLC
Chambersburg PA
CBHW081523220326
41598CB00036B/6304